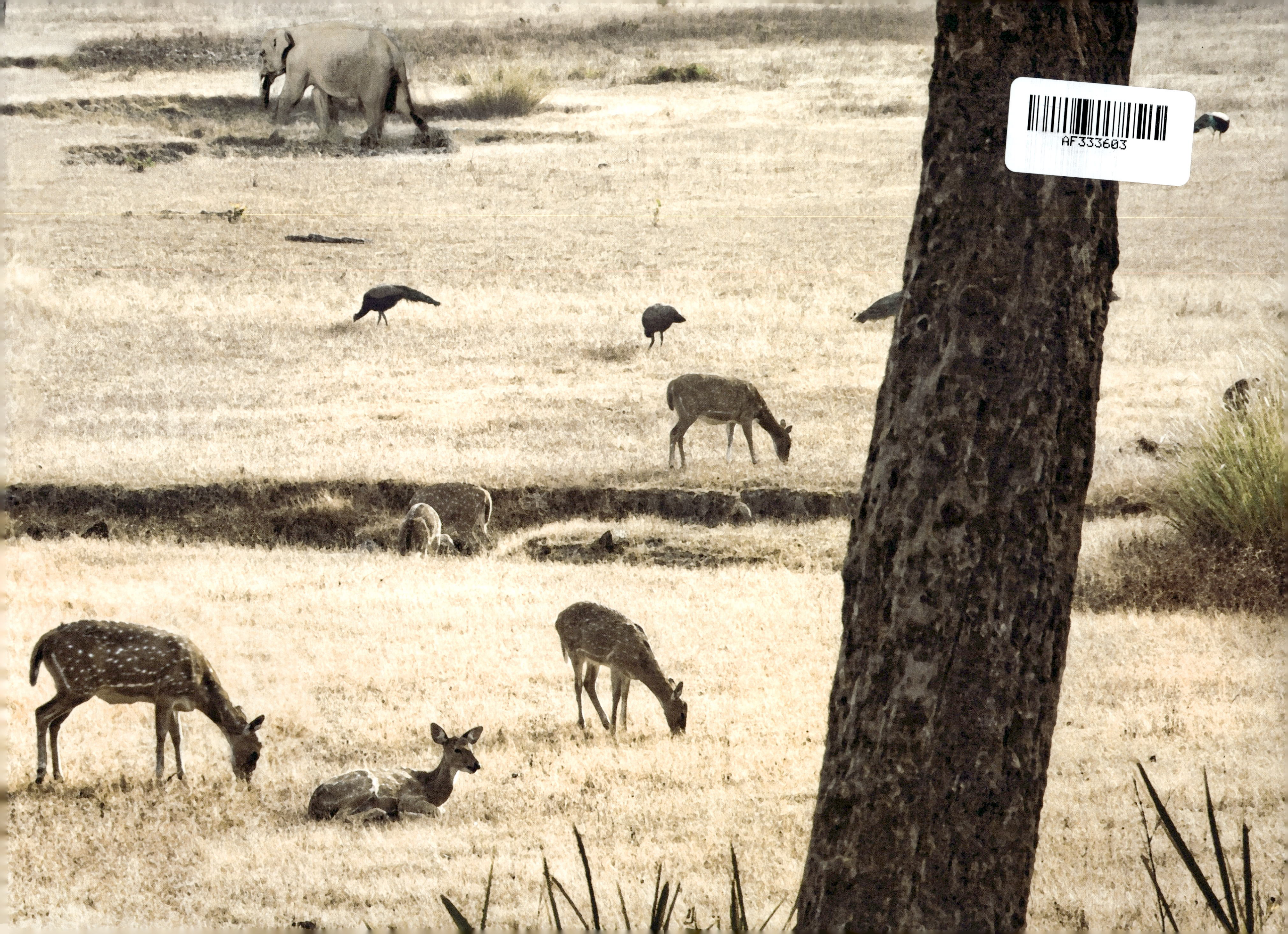

AF333603

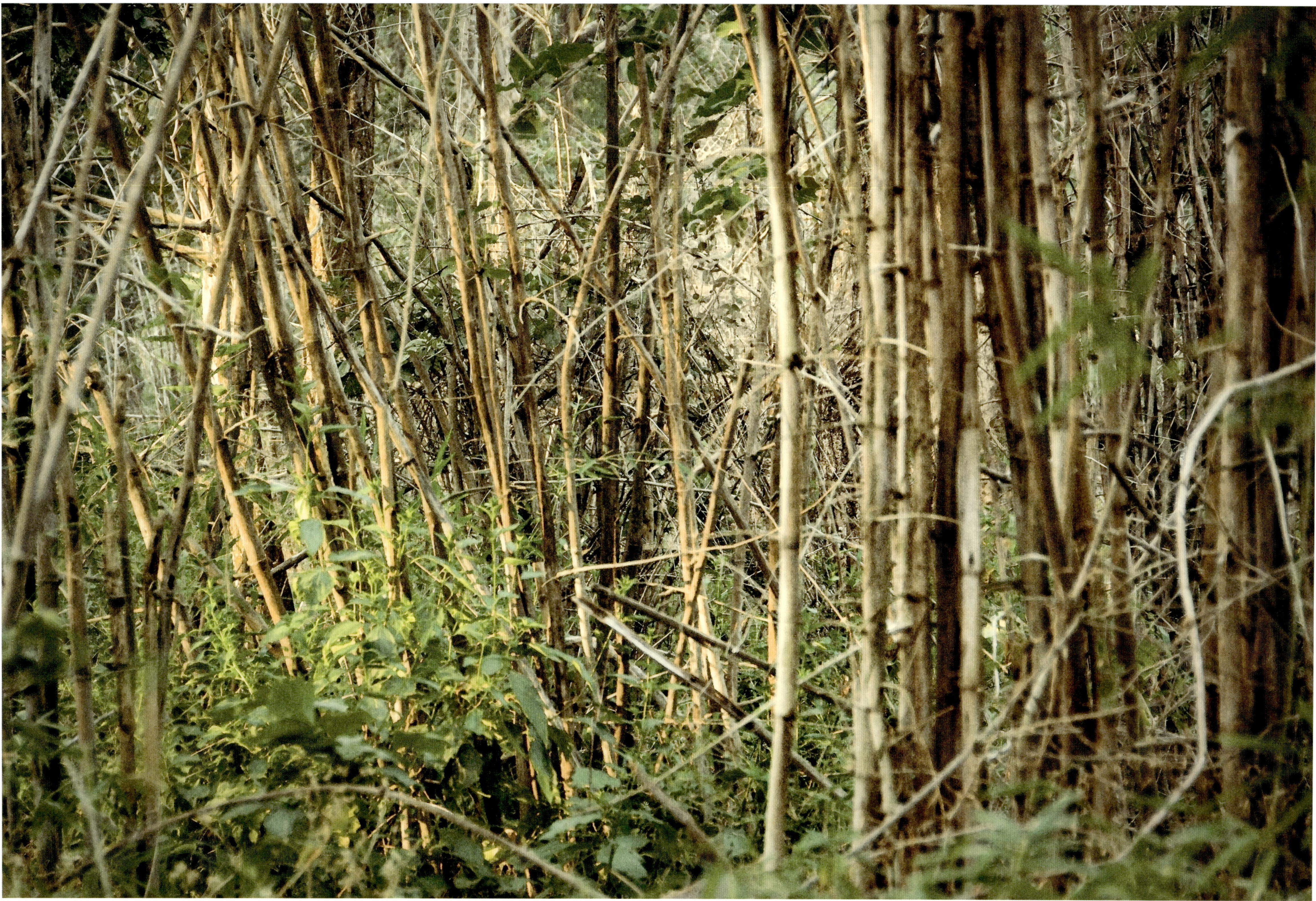

None of the Jungle People like being disturbed.

—Rudyard Kipling, *The Jungle Book*

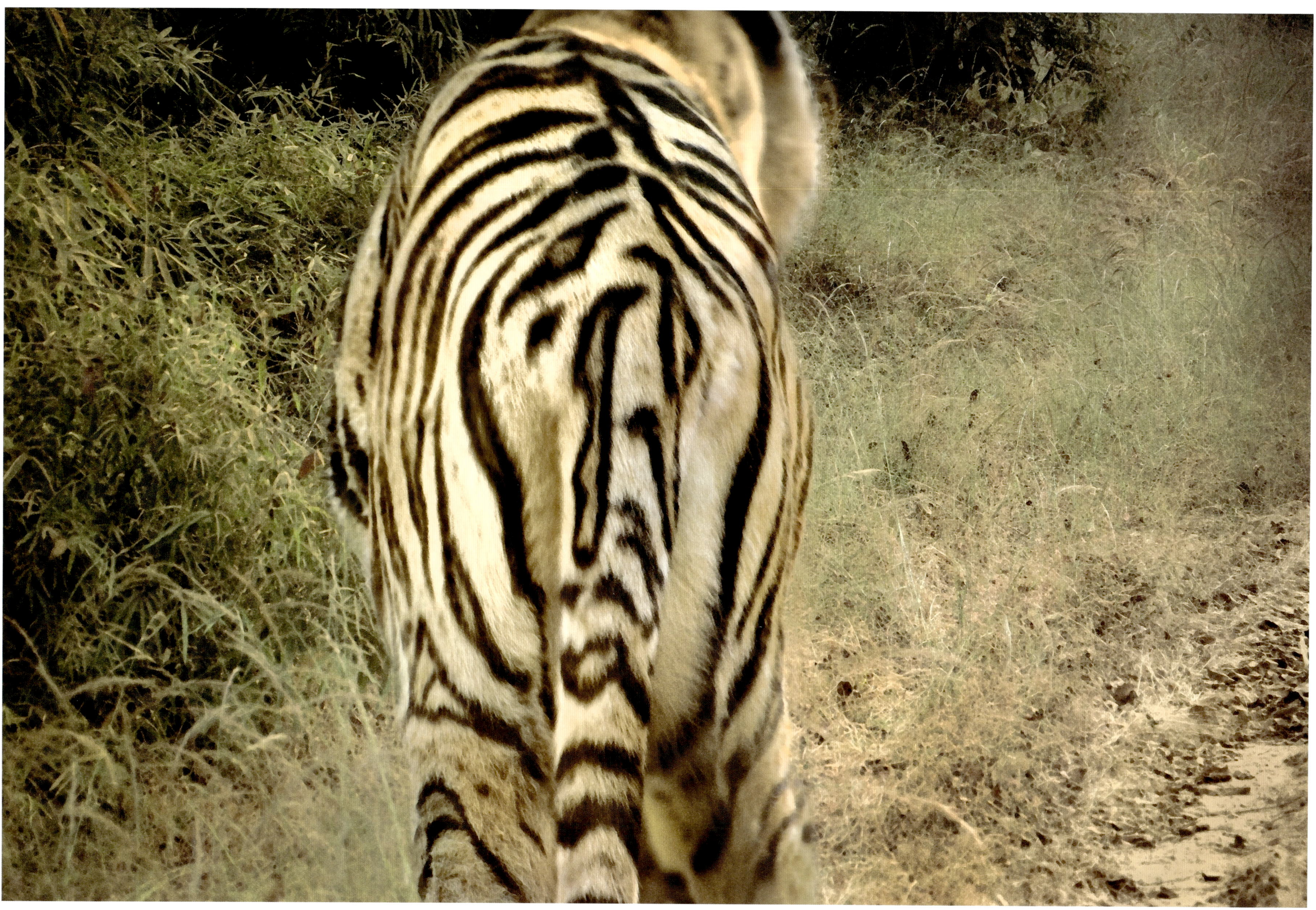

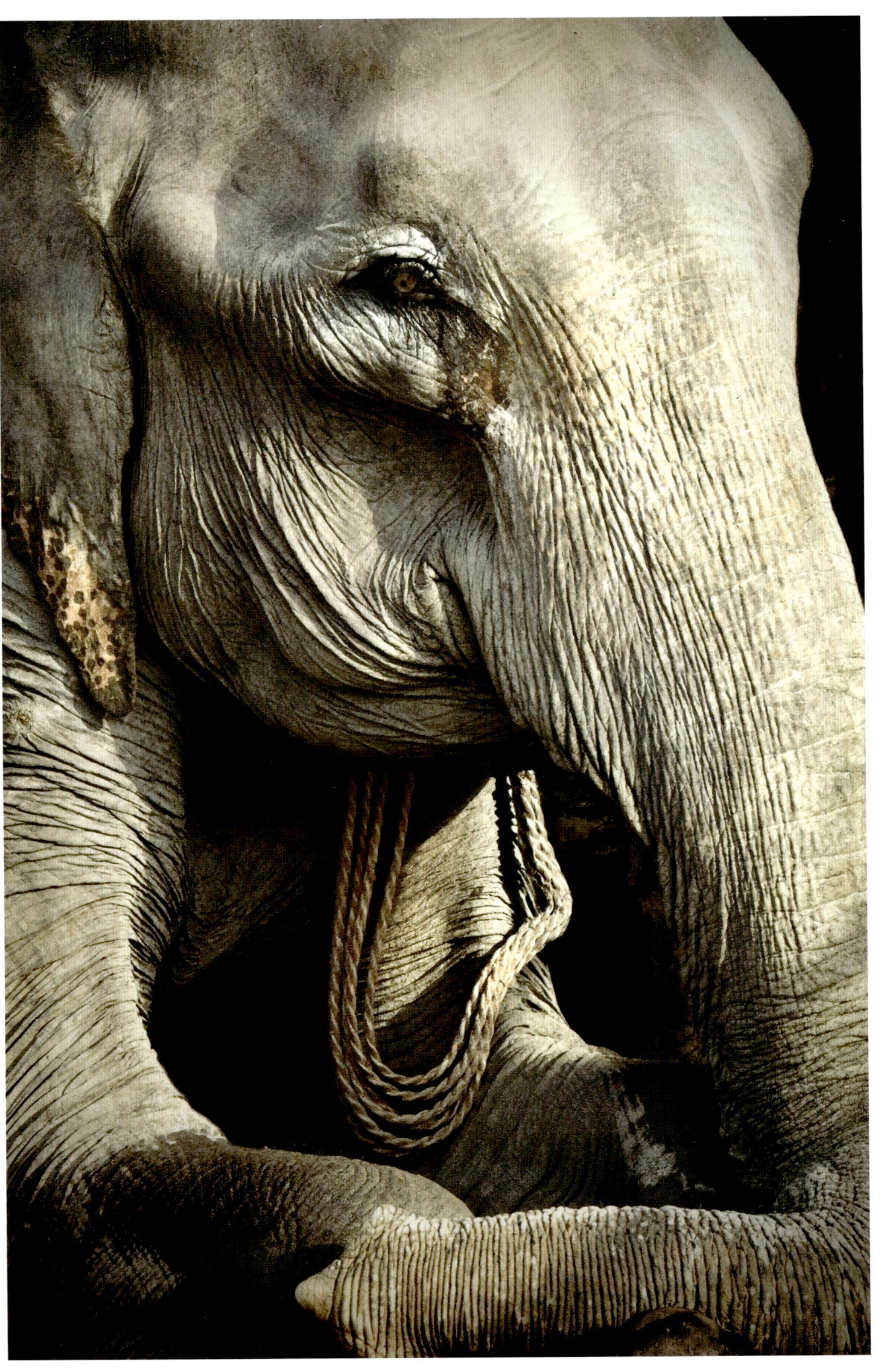

The Jungle
at the Door

A GLIMPSE OF WILD INDIA

BY JOAN MYERS   WITH AN ESSAY BY WILLIAM deBUYS

George F. Thompson Publishing

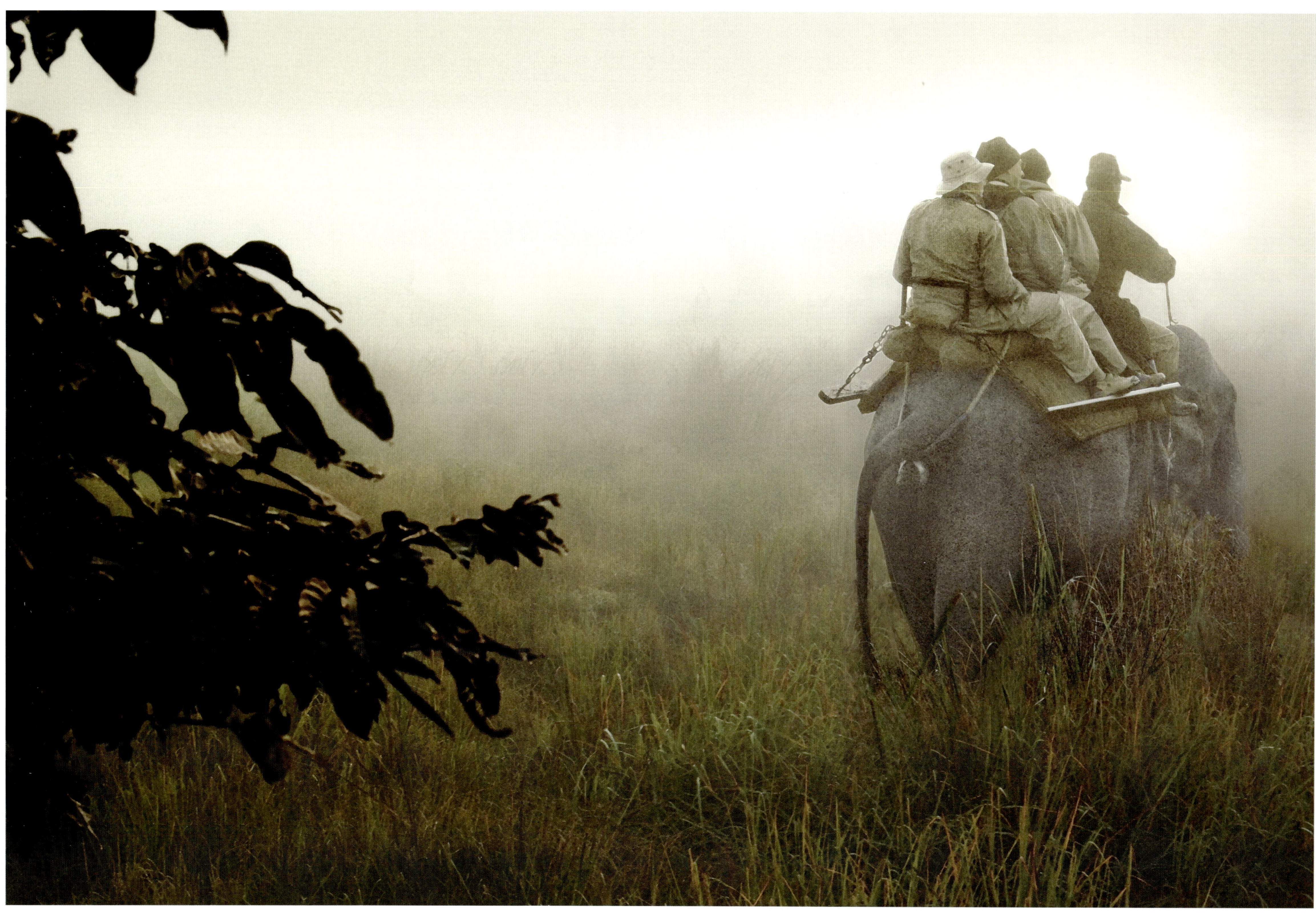

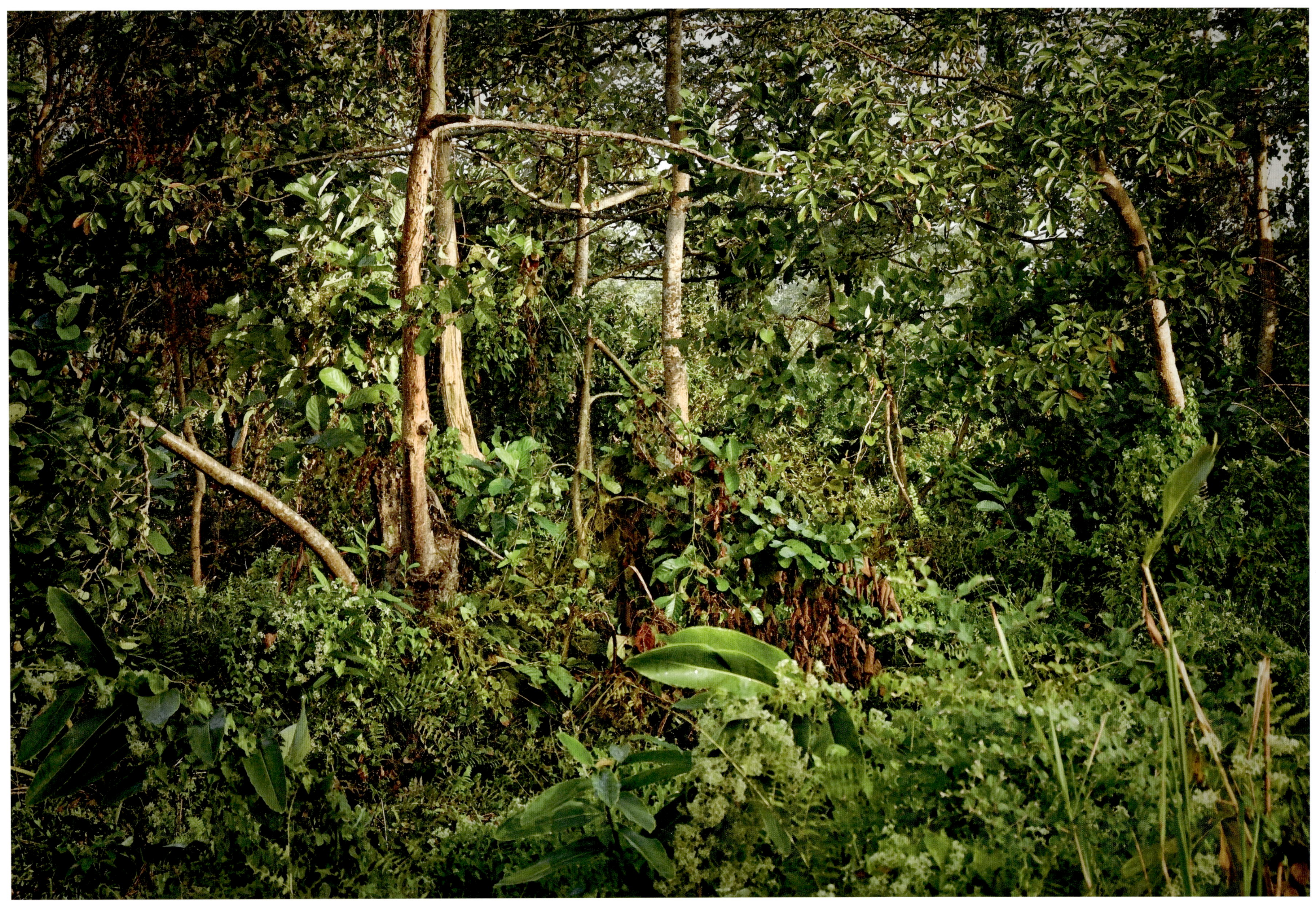

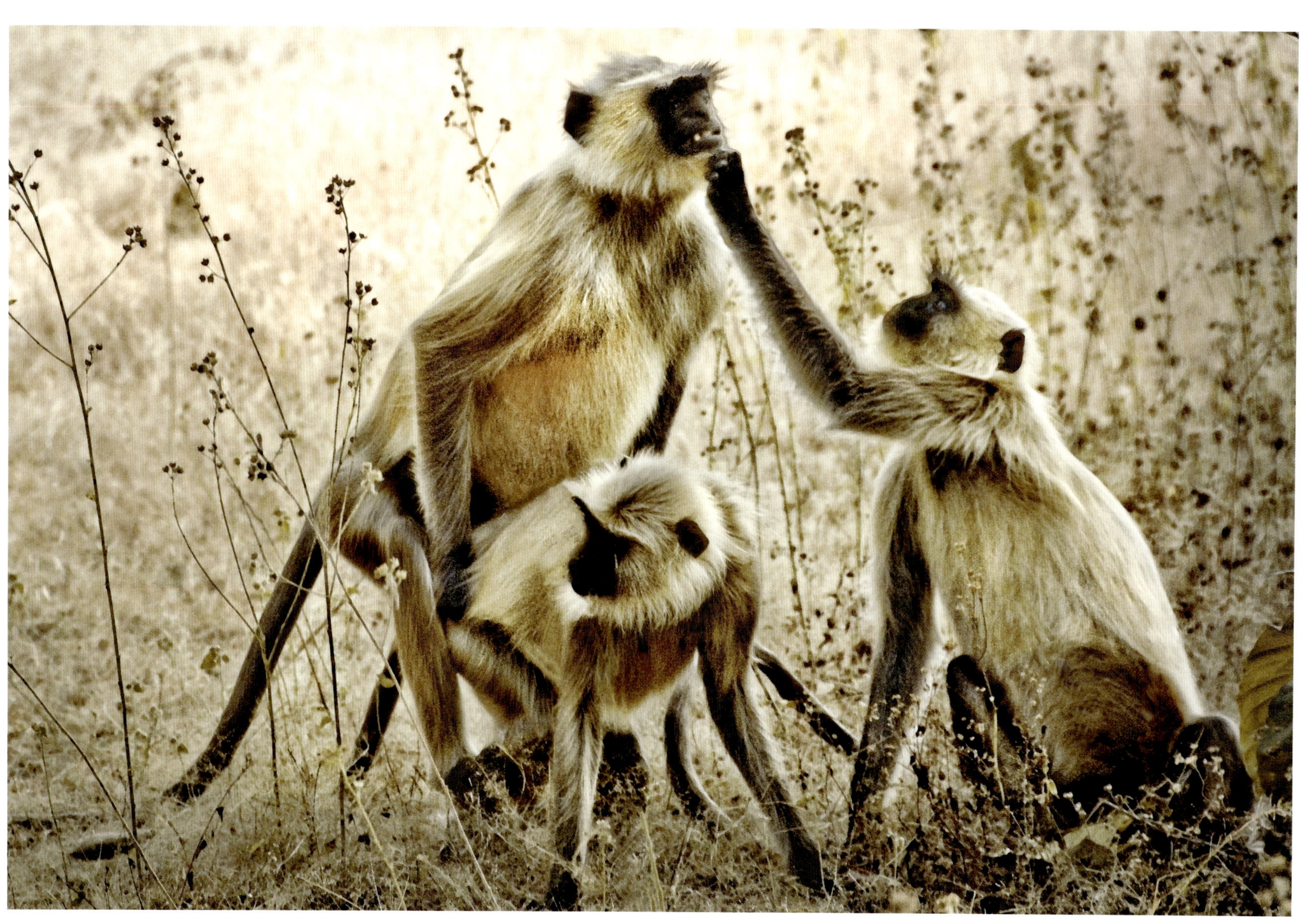

A Glimpse of the Wild

William deBuys

You are walking in a zoo. It is an excellent zoo, with naturalistic habitats—the savannah, for instance, is an archipelago of tawny islands bounded by dry moats. The animals drift in seeming freedom under an open sky. You round a turn on the macadam trail and suddenly before you, on one such island, stand a copse of giraffes, treelike in their height, heads slightly bobbing like wind-stirred boughs. They are taller, more brightly colored—custard yellow with spangles of milk chocolate—and infinitely more improbable than you ever imagined giraffes to be. They move with sleepy grace. One of them turns its head and looks down quizzically. It blinks its long-lashed eyes. Its sides swell with breath. You see the bristly mane, the stubble on the not-quite-horns. The sun, close behind its head, is a little dazzling. The tower of flesh fixes you with a birdlike stare, and you feel unsteady.

In that moment, something shifts. The moats disappear. All barriers fall away. The giraffes, whose number includes a wobbly youngster, lose their zoo-bred complacency and become hyper-alert: they see that the adjacent habitat of the lions is an island no longer. Their new attention is electric: they seem larger still, necks and bodies rocking as they begin to move with speed, their animal architecture so exquisite, so surprising, so sublime. For the first moment in your life, you feel you are really *seeing* giraffes, and the intensity of their beauty is almost unbearable.

You continue through the charmed zoo. You marvel at the china-blue feet of the blue-footed booby, the limpid eyes of the snow leopard, the shaggy beard of the bison. The tail of the argus, cousin to the peacock, is unimaginably long, its patterns like the skin of the most intricately speckled trout.

Fortunately, on this perfect day the predators are not hunting: you are safe. Equally, the other large creatures see no threat in you, so you can linger by the gorillas and admire their prehensile feet. You can stroke the stiff and bottomless white cloak of the polar bear, and, when you later find yourself among the antelopes, you can read, close up, the painted face of the oryx and marvel at the sculpture of its horns. For this once, your mind is not cluttered with borrowed images. You are seeing each animal *de novo*, as though you had never before beheld its likeness in a photograph or painting, let alone in cartoonish advertisements or caricatures. You are eye to eye with the real thing.

But this is just a fantasy. We are rarely so lucky as to see things unfiltered by memory and experience. A lifetime of television programming and issues of *National Geographic* helps us paint our expectations. And, where the documentaries leave off, Tony the Tiger, the Lion King, and Babar begin to nudge our sensibilities more than we care to admit. And if not them, then the Chicago Bears, Panda Express, Curious George, or some other exemplar of the bestiary of popular culture. We rarely slip free of the baggage we carry. As a birder, I catch myself bending the images of the birds I see to match the pictures in my field guides. I can't seem to help it.

Once, though, I blundered into a kind of freedom. It was thirty-five years ago, but the memory remains vivid. I was at home, in mountain country. A wet snow had fallen, and a heavy, opaque sky blotted the sun. No shadows hinted the passage of time. No wind disturbed the silence. I was prowling around, purposeless, binoculars in hand. I glimpsed a bird darting into the snow-covered canopy of a low juniper. I circled round to approach from the opposite side. A jiggle of twigs told me the bird was still there. I crept nearer, but I could not see it; the foliage was too dense. I ducked under the first branches and crouched within.

Inside was dim space, leaving the outside world muffled, suddenly distant. The bird perched on a branch only three feet away and watched me with a cold, black eye. I did not recognize the species. It was new—a find! I held as still as I could, on bent knees that wanted to tremble, and I marveled that the bird did not fly off. Perhaps the strange, murky day had cast a spell on it, as it had on me. Close up, the feathers on the back of the bird gleamed like satin. They were a hue married from gunmetal and treebark, a shade of dusk, for which I have no word. The beak was the dun of deer horn, and a ring of white encircled the eye. From the throat down to the downy feathers of the belly, the bird seemed nearly to glow. Its breast was the orange fading to russet that you see in the last moments of a tropical sunset, the color lurid yet delicious, like the flesh of exotic fruit. It was a color that seized the eye and would not let it go. It was a color you would paint a portion of your world with, if only you could replicate it.

The bird stared at me and I at it. It twitched its head and hopped one branch farther away, the russet of its underside still shining in the half-light. A minute passed, and I realized I had ceased to care about identifying its species. The bird was so profoundly *there*, so close and alive. Strange to relate, but I began to feel the hardness and completeness of the bird's separate being. The feeling was strong because I could not place the bird in a category—it was neither sparrow, finch, nor dove. I had no name and no template for it, and so it was just *bird*, fresh from creation, and marvelous.

Then, with a cackle, the bird burst from its perch and flew away. The sound and movement woke me from my trance and I realized I knew very well what sort of bird it was. It was a robin, *Turdus migratorius*, the commonest of the common. I felt embarrassed and thick-witted—I'd become a satire of myself! But then I recalled how striking and vivid the bird had looked and how it felt to have that black eye bore into me. Never before had I sensed the presence of another creature so strongly, and, I am sorry to say, I rarely have again. In that moment, an ordinary robin seemed a greater miracle of nature than even a Bengal tiger, "burning bright, in the forest of the night."

Wildlife photography, including video and film, brings such miracles to our living rooms and offices, our cinemas, televisions and computers. It greatly extends our natural vision of the wild, sometimes at the risk of supplanting it. It has become an art of its own, and its foremost practitioners are legendary for their patience, daring, and technical proficiency. For example, consider the challenges that had to be overcome in order to capture one of the most remarkable recent achievements of the genre: start-to-finish footage of a snow leopard hunting and nearly taking a juvenile markhor (a species of wild goat) in the Karakoram Mountains of Pakistan. First, the team had to struggle with political and military permissions, then with travel of the most arduous kind. The land is supremely hard, the elements severe. Finally encamped, the photographer and goodness knows how many assistants rise in bitter cold before dawn, take cover in a camera blind on the face of a cliff above a river, and spend the day scanning the cliff face on the opposite side of the river. As patient as the predator they seek, their lying in wait goes on for days, and nothing happens—the bored sun moves in its slow arc, and the wind blows.

But they cannot surrender to boredom because, if anything is ever to happen, it will happen with terrifying speed, and one has to be ready. And so one day a band of markhor straggles into view, clambering improbably along the near-vertical face of the cliff. They are shaggy goats with spiral horns, among the rarest grazing animals on Earth. And *look at that,* a snow leopard stalks from above! With lenses suitable for advanced astronomy, the photographer zooms in until the leopard fills his viewfinder. Hundreds of meters of distance—the width of the canyon—fall away. You see the leopard's broad silent paws grasp the rock; you see how the sinuous elegant tail counterbalances the forward tilt of the body. And then the downslope charge at the juvenile. The chase, the clatter of rocks, the ferocious bite on the rump of the juvenile. The two creatures, now joined, slide together down the cliff. The leopard attempts to hold the juvenile back, to check their downward plunge, to secure both of them against the rush of gravity. But in vain. The juvenile breaks free, and it leaps! It bounds into the frigid waters of the rushing river, undoubtedly to drown. The leopard watches it go.

The camera gets it all. The mortal sequence is recorded, preserved forever. It is footage as exotic and as beyond the reach of normal senses as radio telemetry from a space probe. It is soon to be replayed, and replayed and replayed in parlors planet-wide.

There is another kind of wildlife photography, a humbler, more modest kind that does not endow us with the vision of Superman. It more nearly replicates the act of seeing with human eyes. It lacks omniscience. Although it cannot show us the mysteries of unreachable worlds, it is no less important. Its fleeting moments are blurred, like memory, and obscured by mist. Its depth of field is split, like our natural powers of attention: some things are in focus, others not. And it is emotional, not cool and detached: *look*, the bamboo is frighteningly dense: what's in there? And *look there*, it's the big cat but screened by leaves, or *there*, half hidden by a tree. You never see all of it, and then, *nothing*, it is gone. But *here*, do you see? This big deer, this sambar, is the tiger's kill. Let's get out of here.

In some places where tigers live it is deemed unwise to say the name of the beast because to utter "Tiger" is to call it, to summon it. Summoning a tiger is not a good idea if you are a skinny, unarmed earthbound human. Some wildlife photography, like the sequence of the snow leopard and the markhor, takes us deep into the otherwise alien world in which wildlife live. We become as gods, seeing without limits. Other wildlife photography places us *in relation* to nature's creatures. In its perspective, we are merely primates, oftentimes vulnerable, sharing our Earth with wild things and limited in our knowing them both by our finite senses and our trepidation. This quick, brilliant, impressionistic collection by Joan Myers fits the latter description. It has an Edward-R-Murrow-sense of *you are there.* And, just as important, it puts you there *as you are,* traveling light. You are a fellow creature, not the master of creation. For once, you are restricted to, and blessed with, a kind of equality with wildlife. It is the right starting point for seeing them on their terms.

Demographers tell us that India will eventually pass China as the most populous nation on Earth. Forty-one people are born there every minute. Tick tock. The pressure to find space and resources for the country's swelling human population dwarfs most other concerns, and yet India is vital to the future of the planet's wildlife. Its remaining natural habitats are among the last, best places for an entire arkful of the world's most spectacular and inspiring animals, and yet settlement runs right to the edge and often into the country's wildest protected areas. There is often no zone of transition, no buffer. The jungle is jungle, and then it simply stops—sometimes literally at someone's door. And the people who live behind such doors, live in constant tension with the creatures of the wild, preeminent among which is the tiger.

Tigers are extinct in ninety-three percent of their historical habitat. Their wild populations in China, Laos, Myanmar, and Vietnam are at best in the low two-digits. Of the 3,200 wild tigers believed still to exist, approximately half survive in India. Consequently, the conservation of tigers in India must be central to any plan to save the species from extinction in the wild.

The threats are multiple. They include habitat fragmentation and degradation, encroachment on protected areas, and depletion of prey species (which, if it does not result in starvation or reproductive failure, leads to predation on domestic livestock, with ultimately disastrous consequences for the tiger). Most pernicious of all, however, is poaching for the international trade in animal parts. In Traditional Chinese Medicine (TCM), tiger bone is considered a medicine of incalculable power. It is penicillin, interferon, and Viagra rolled into one. As economic prosperity increases in China, South Korea, Vietnam, and other Asian "tigers," as these manufacturing powerhouses are known, many more millions of people are acquiring the means to pay for high-priced treatments. And the treatments grow more extravagantly expensive as the source animals—turtles, pangolins, snakes, bears, you name it and, of course, tigers—grow rarer. But demand does not slacken. It only grows, even to the point that it spills over to non-target species. Bones from lions, poached in Africa, are now being sold as tiger counterfeit in East Asia.

The plight of rhinos is similar. The last Javan rhino in Vietnam died of wounds received from a poacher in 2010. In 2011, the western black rhino of Africa was declared extinct, another victim of poaching for the wildlife trade. Even taxidermied animals have not been safe. Thieves have broken into dozens of museums and natural history collections in Europe and made off with rhino horns, which are then smuggled into the TCM trade. Increasingly, the conservation of the world's most charismatic species is being won only at the point of a gun. Animals like tigers, rhinos, lions, elephants, and many more, all with prices on their heads, require the protection of armed guards, squads of militia, rangers, and wardens on constant patrol.

It is not hard to imagine a dystopic future in which the great forests of the world are empty of the species that earlier generations referred to as their "royalty." No more King of the Forest. Bert Lahr in *The Wizard of Oz* would be an amusing anachronism and *The Jungle Book* a merely mythological compilation. Children would think of *T. rex* and elephants as co-occupants of a distant Lost World, accessible only in dreams and fiction. Perhaps sports teams, having exhausted the possibilities among viable higher life forms, will take to naming themselves for invertebrates. Come to think of it, with the NBA's New Orleans Hornets, we're already there.

The antidote to such an entropic, depleted future is, first of all, the celebration of the marvelous creatures still present in our world. The greatest tribute we can give them is to refresh our awe for their magnificence and to see them clearly, originally, *as they are.* And then, having seen them and having become infected by their magic, to do something, to do a lot, alone and together, to save them a place on our shared blue planet. They and the rest of Earth's biota are our only companions within the vastness of the universe. Without them our loneliness would stretch to infinity.

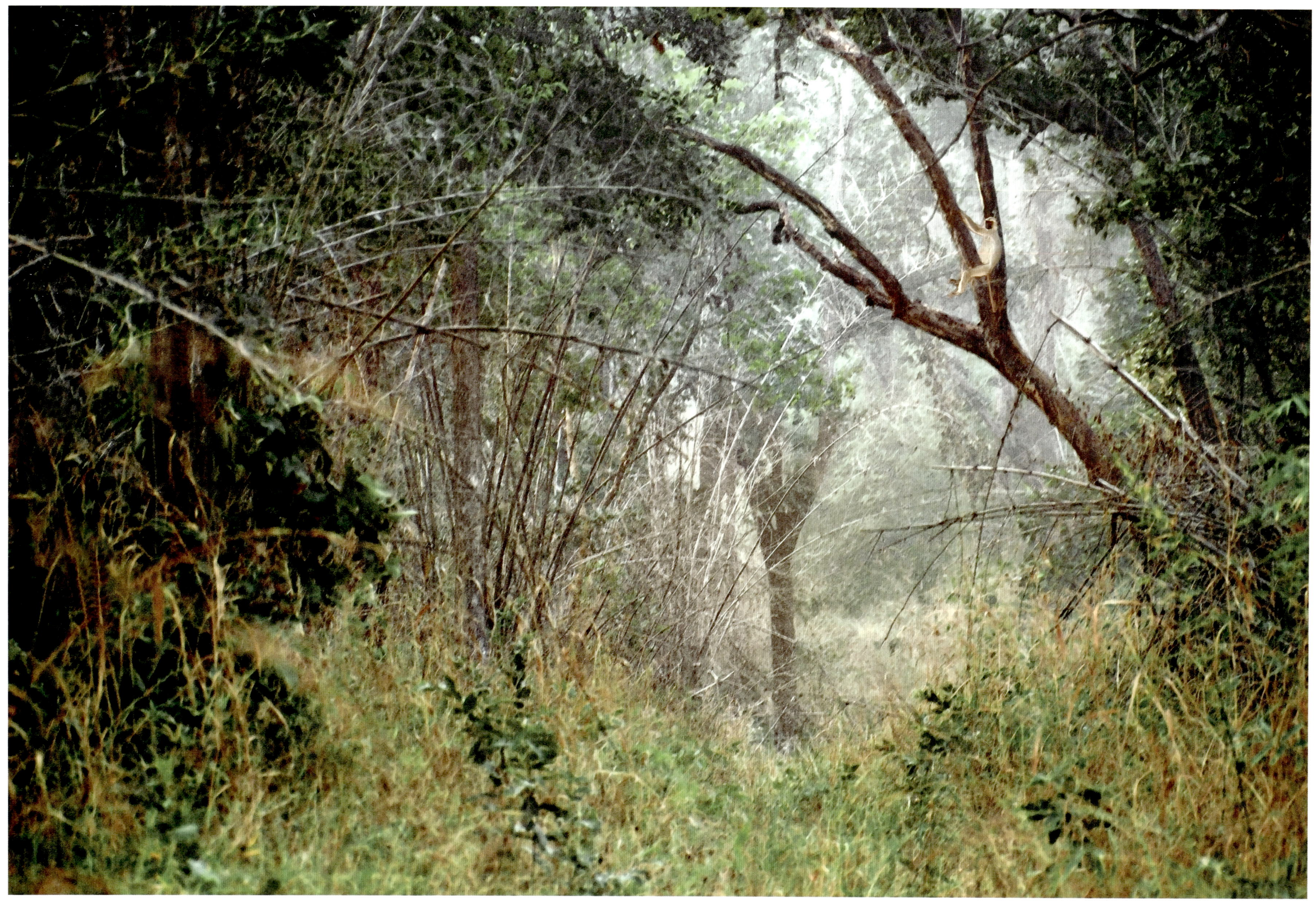

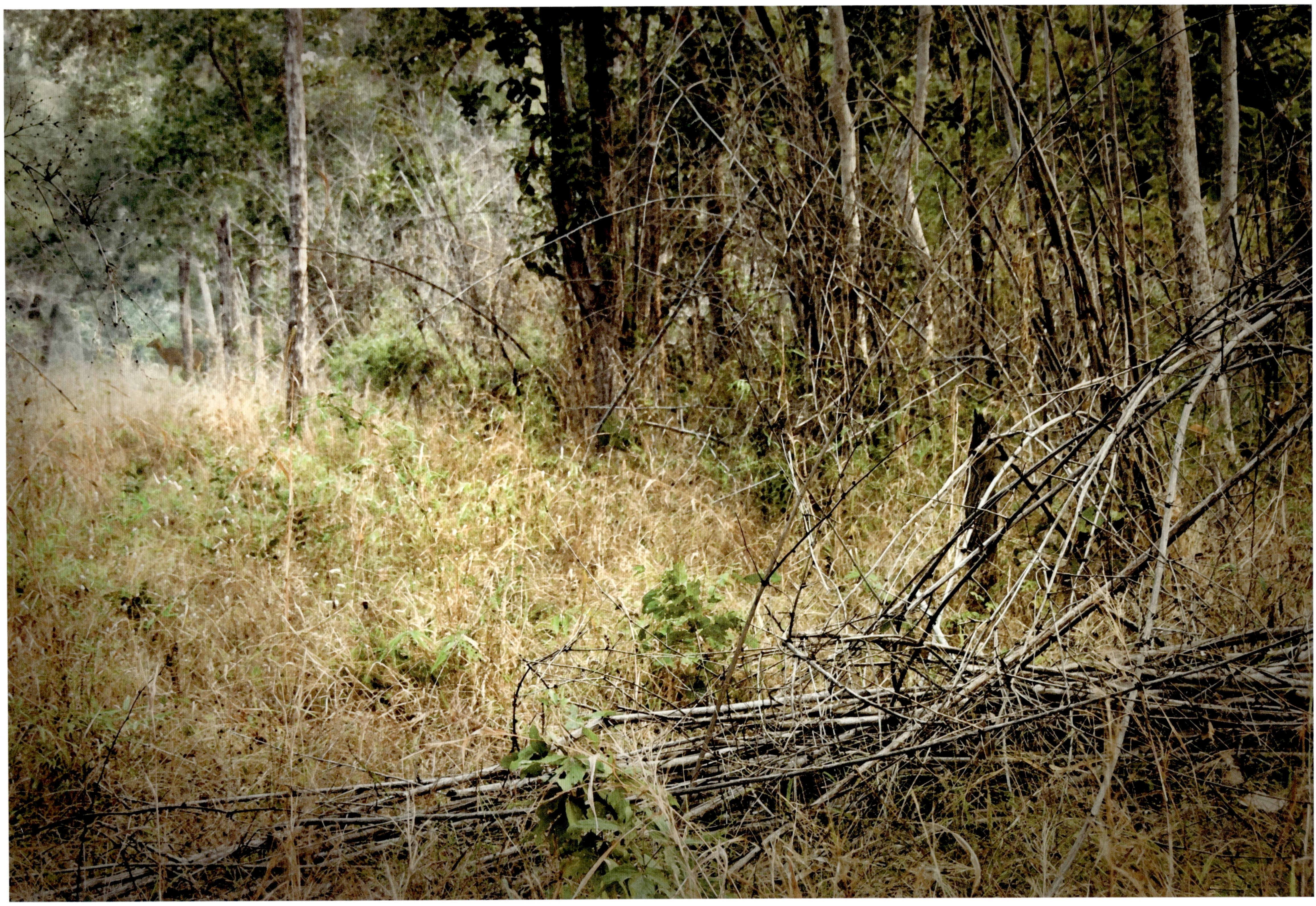

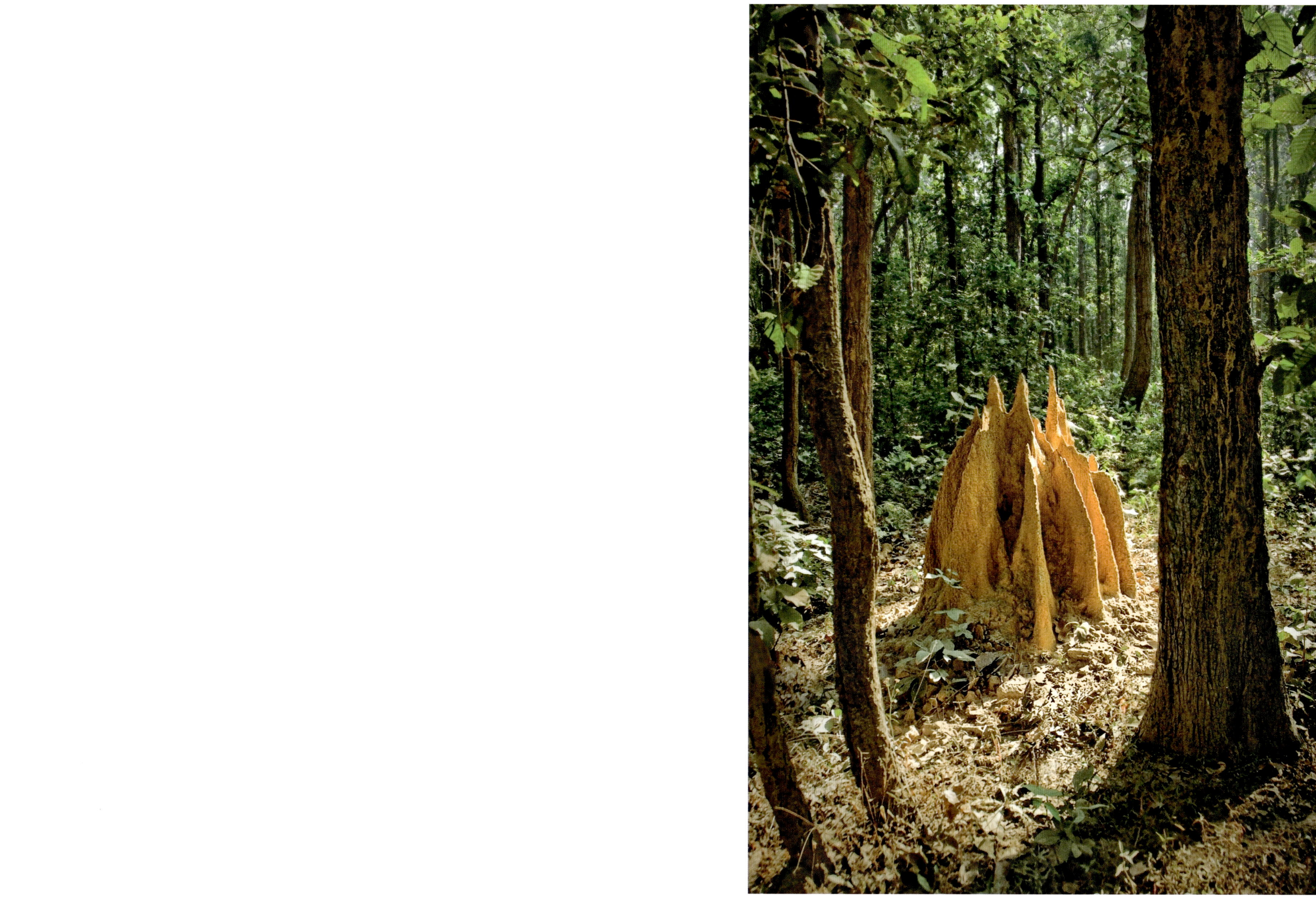

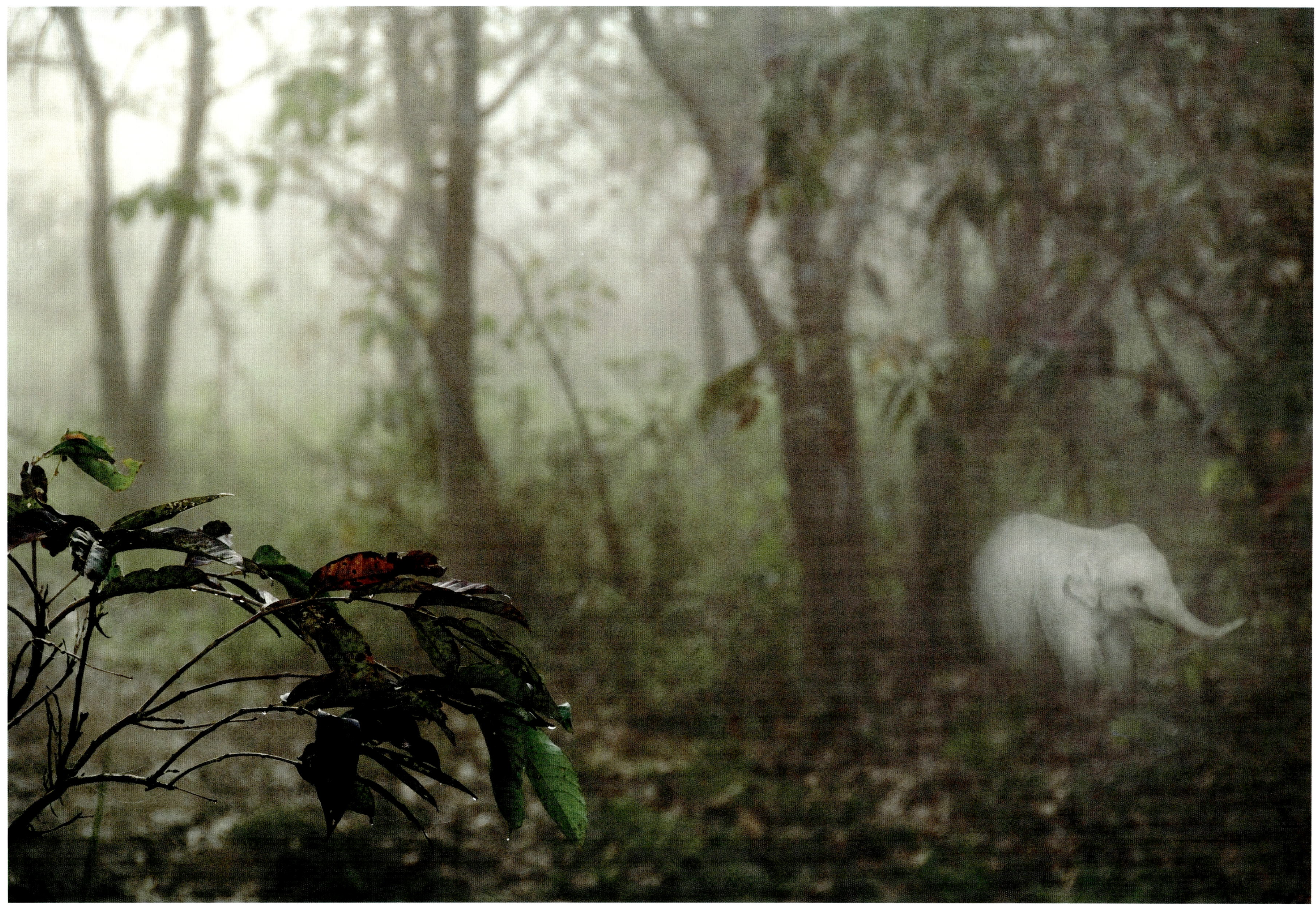

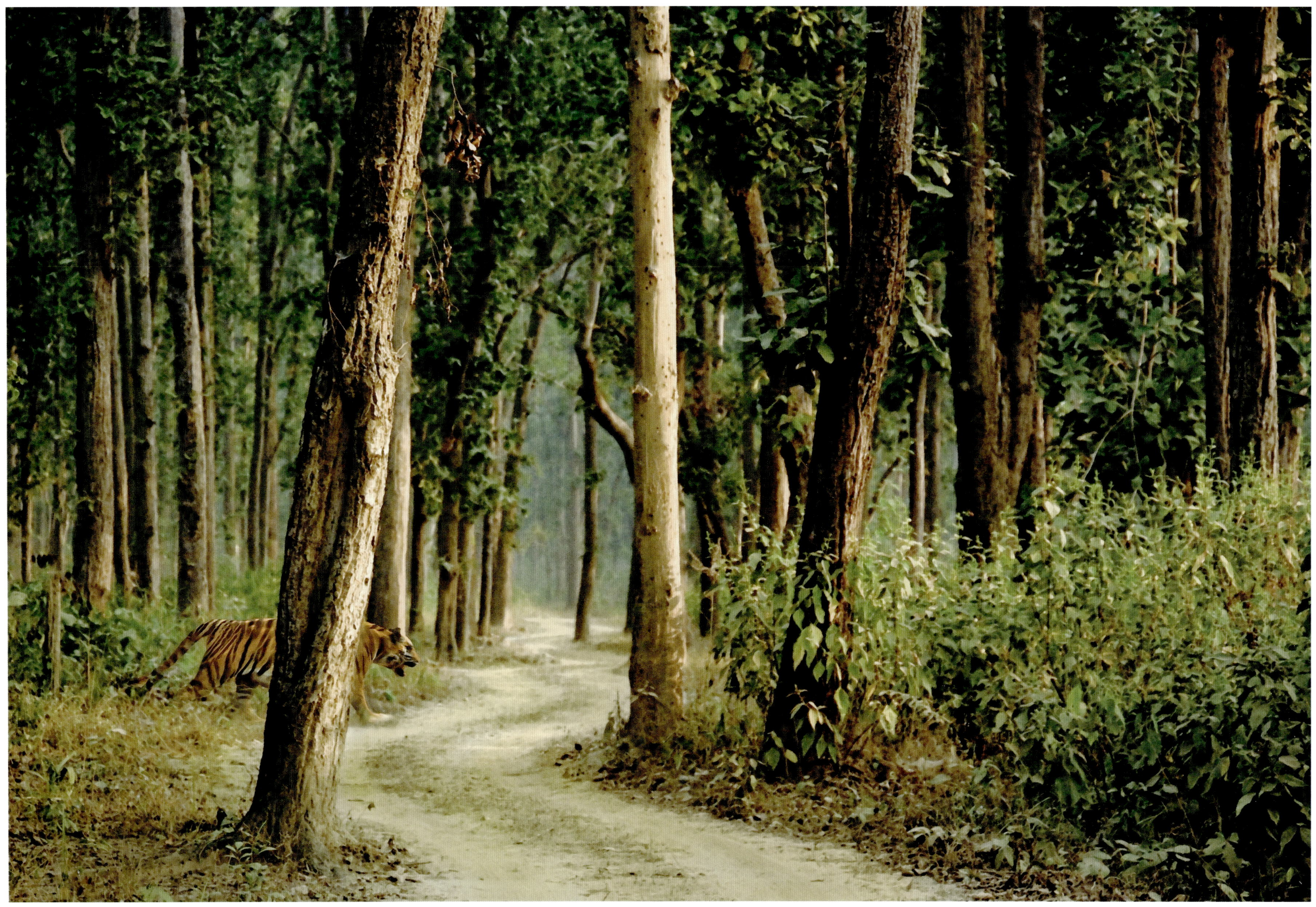

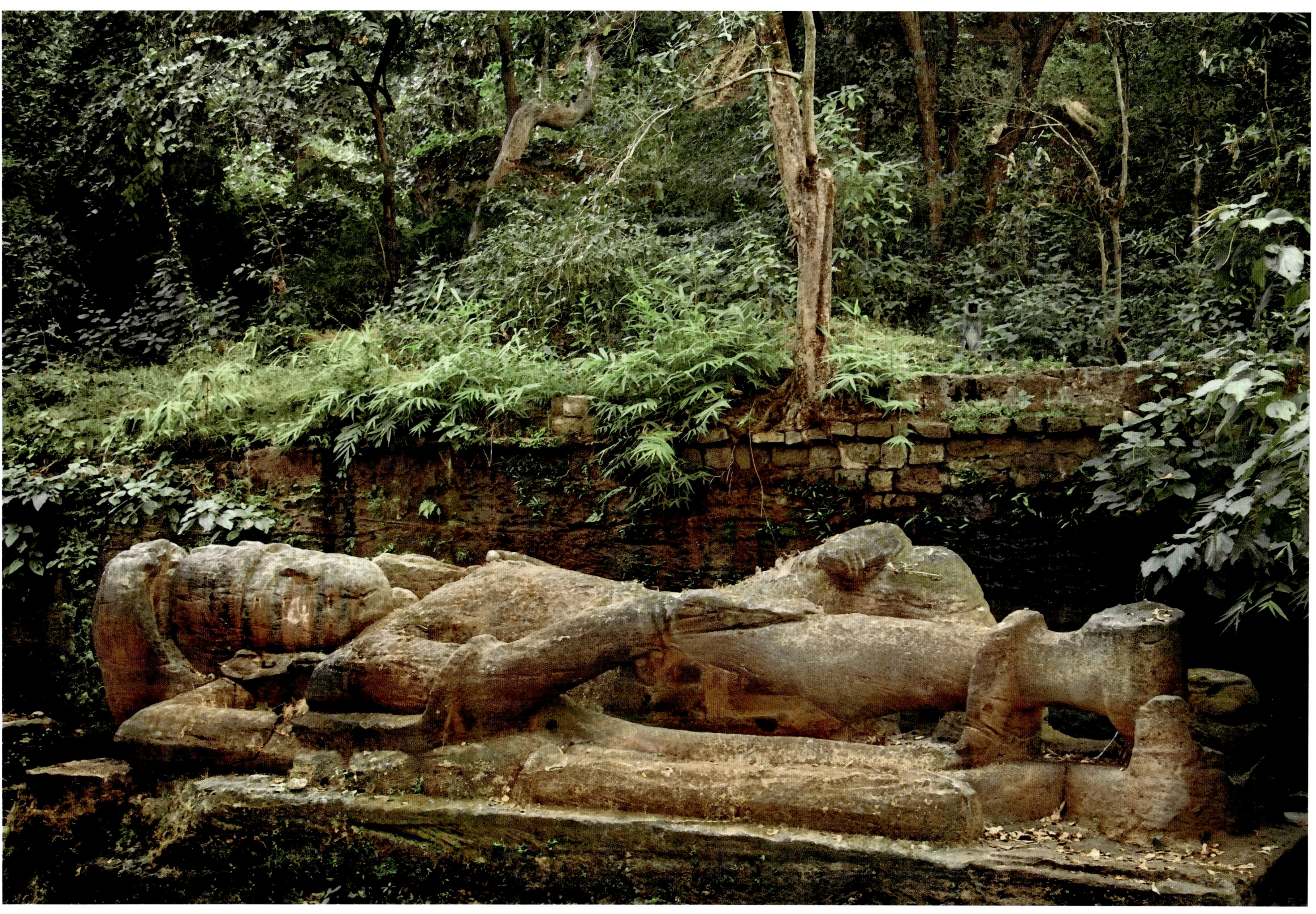

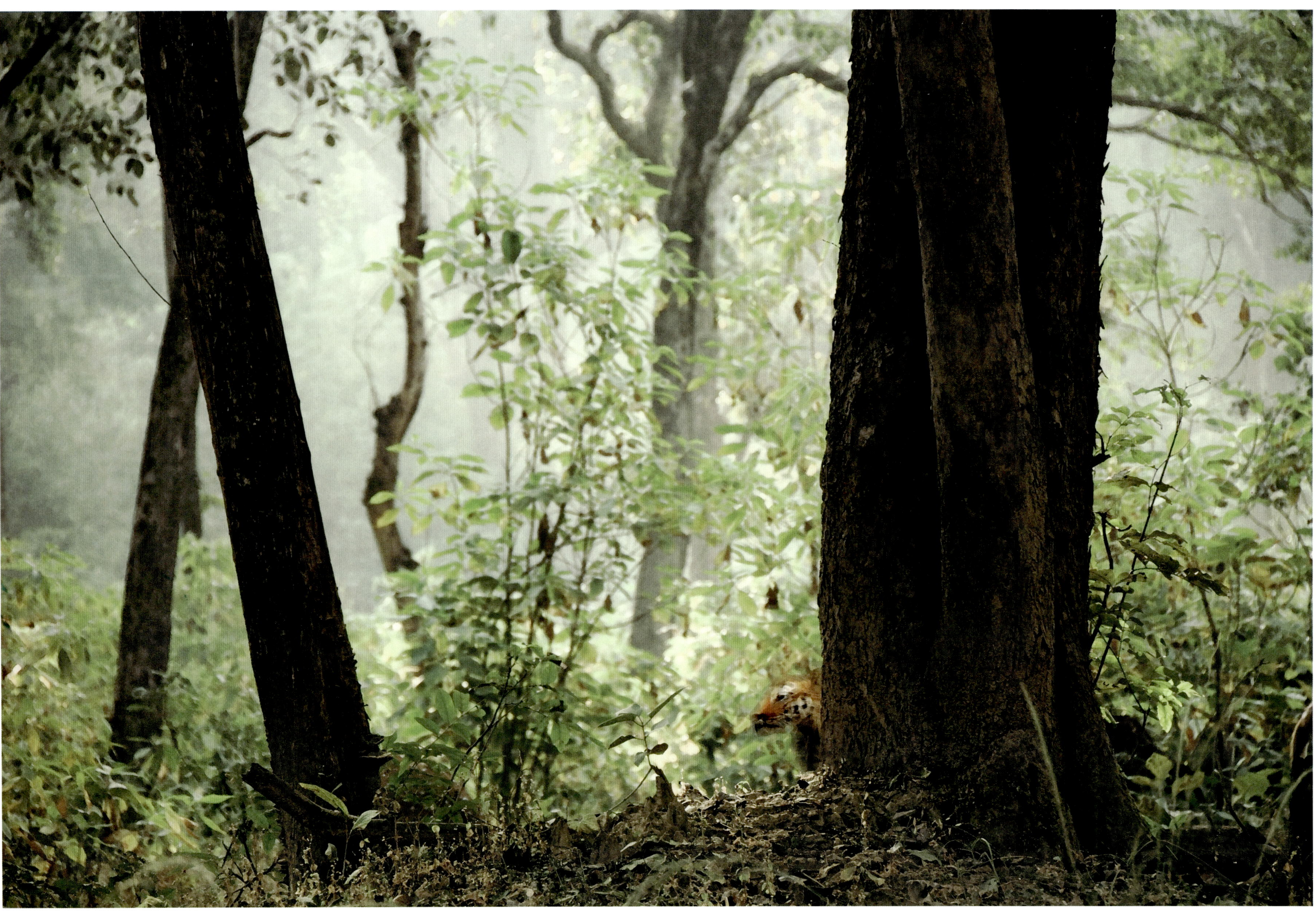

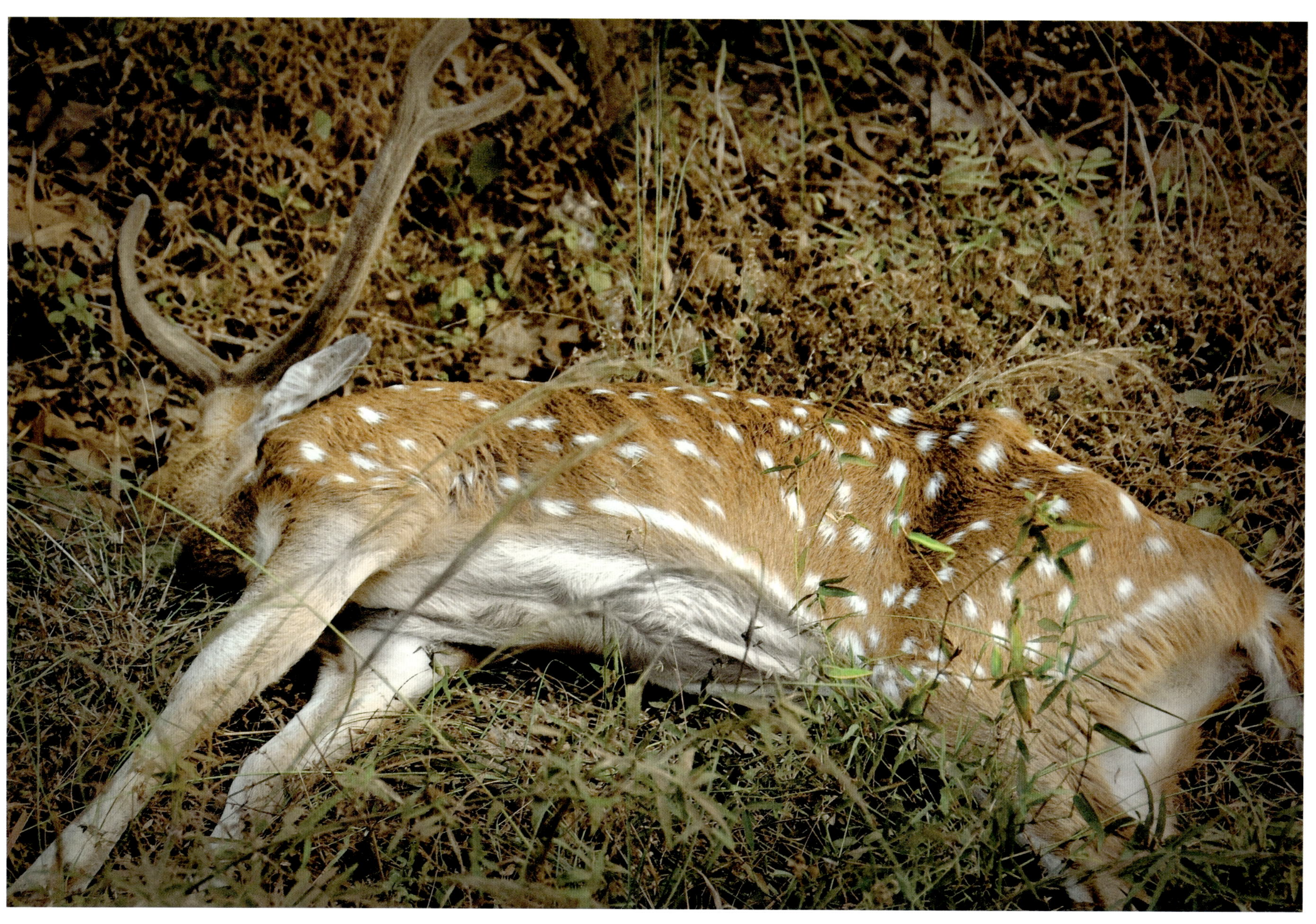

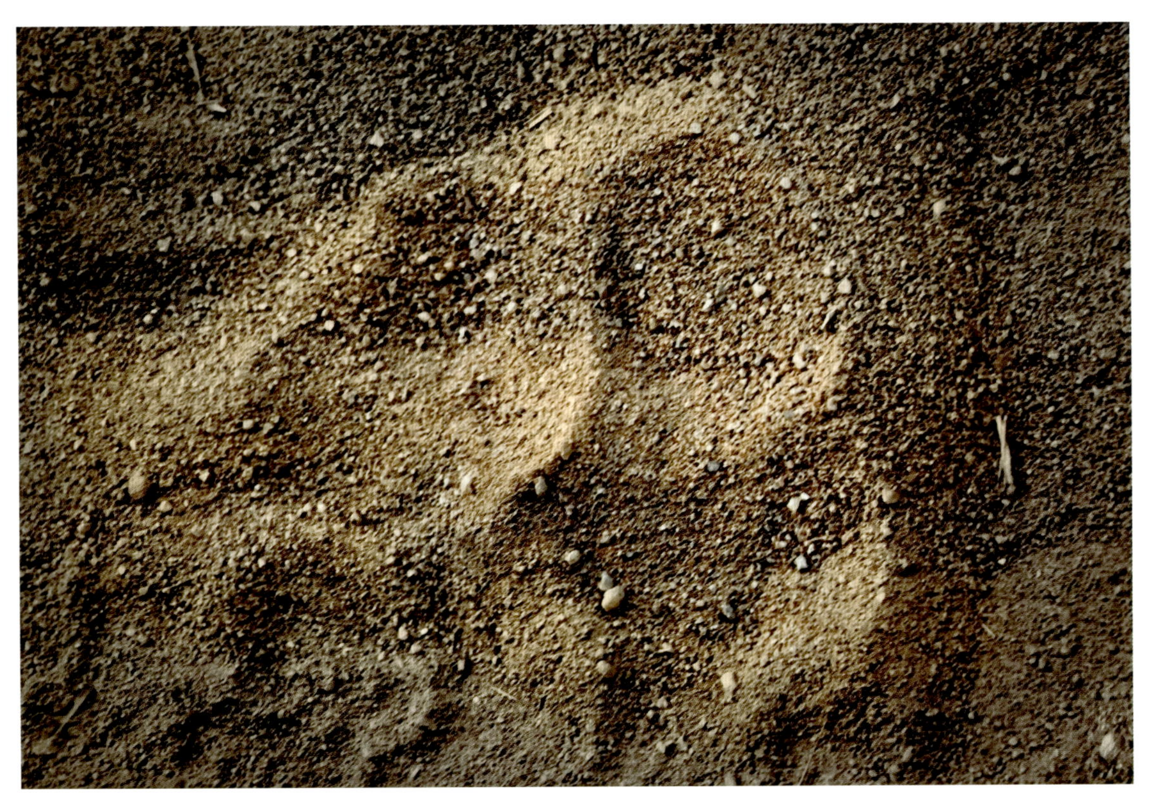
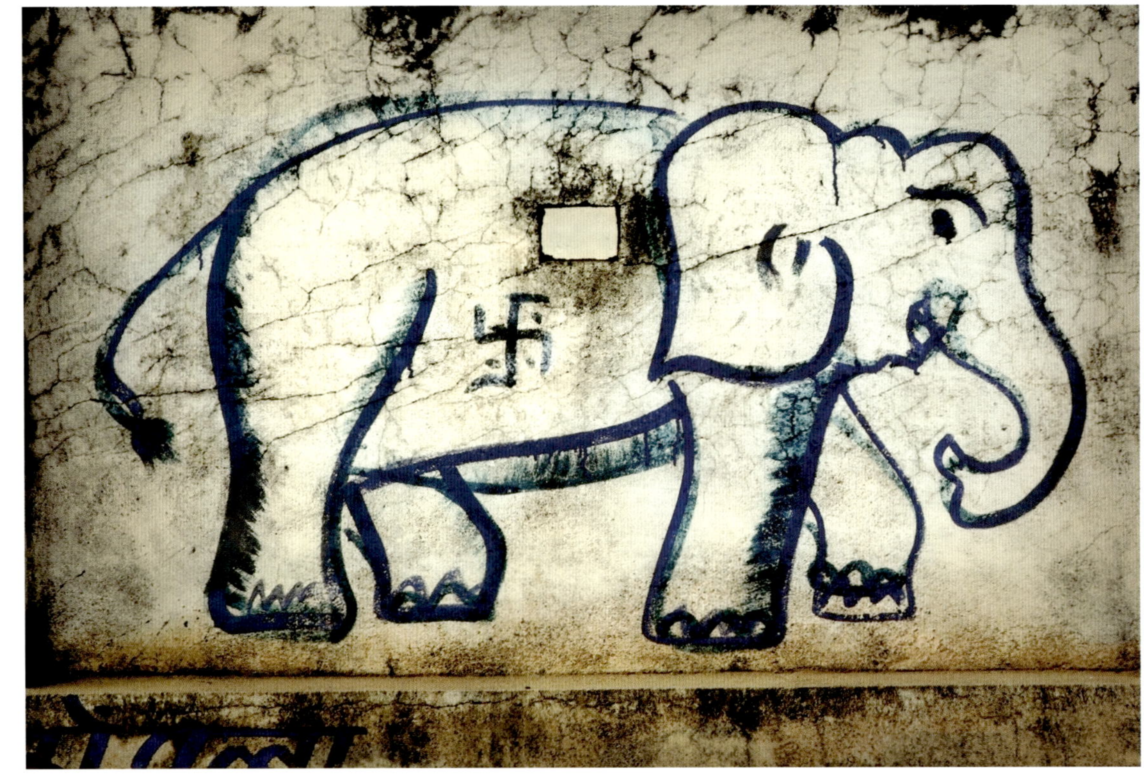

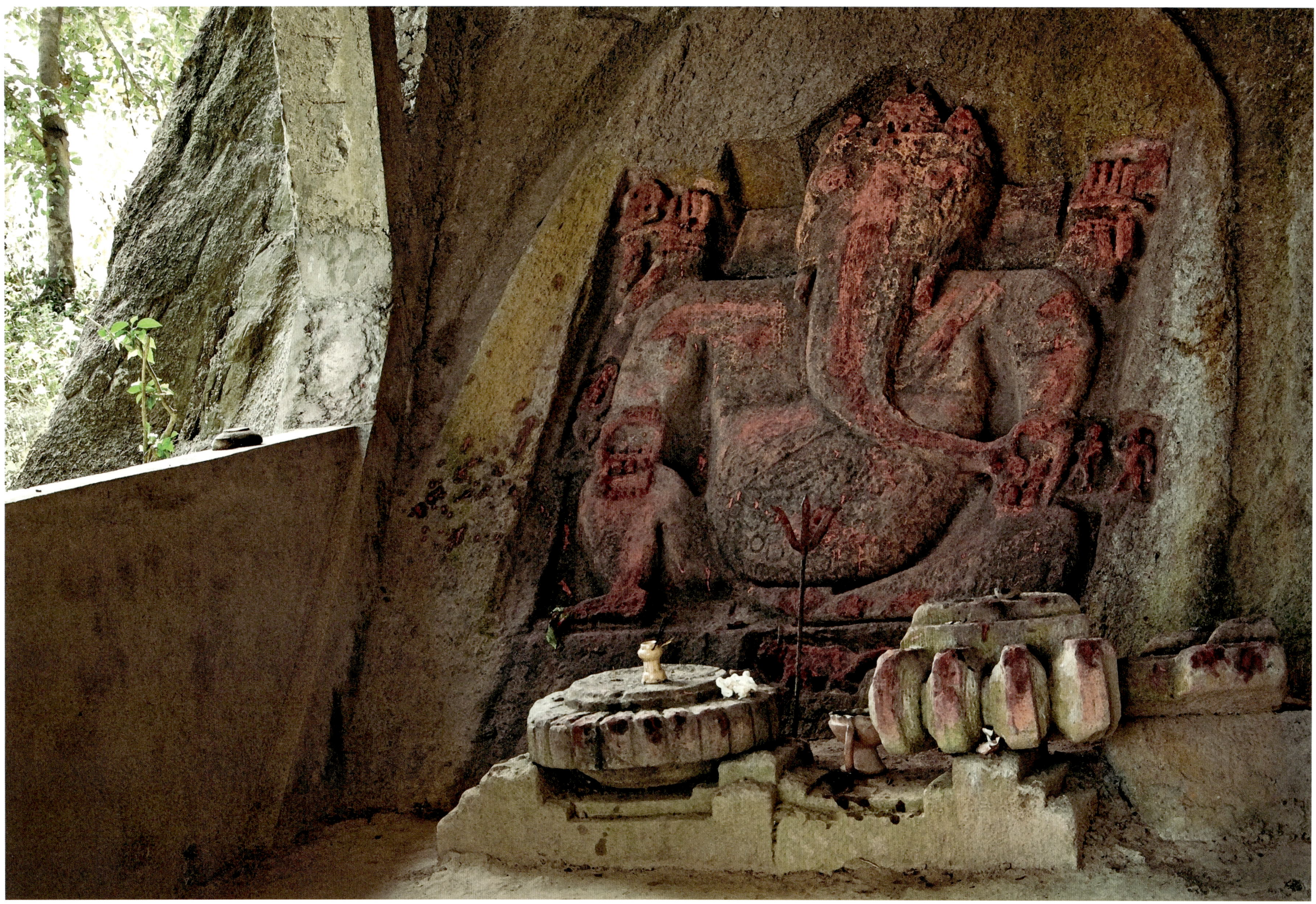

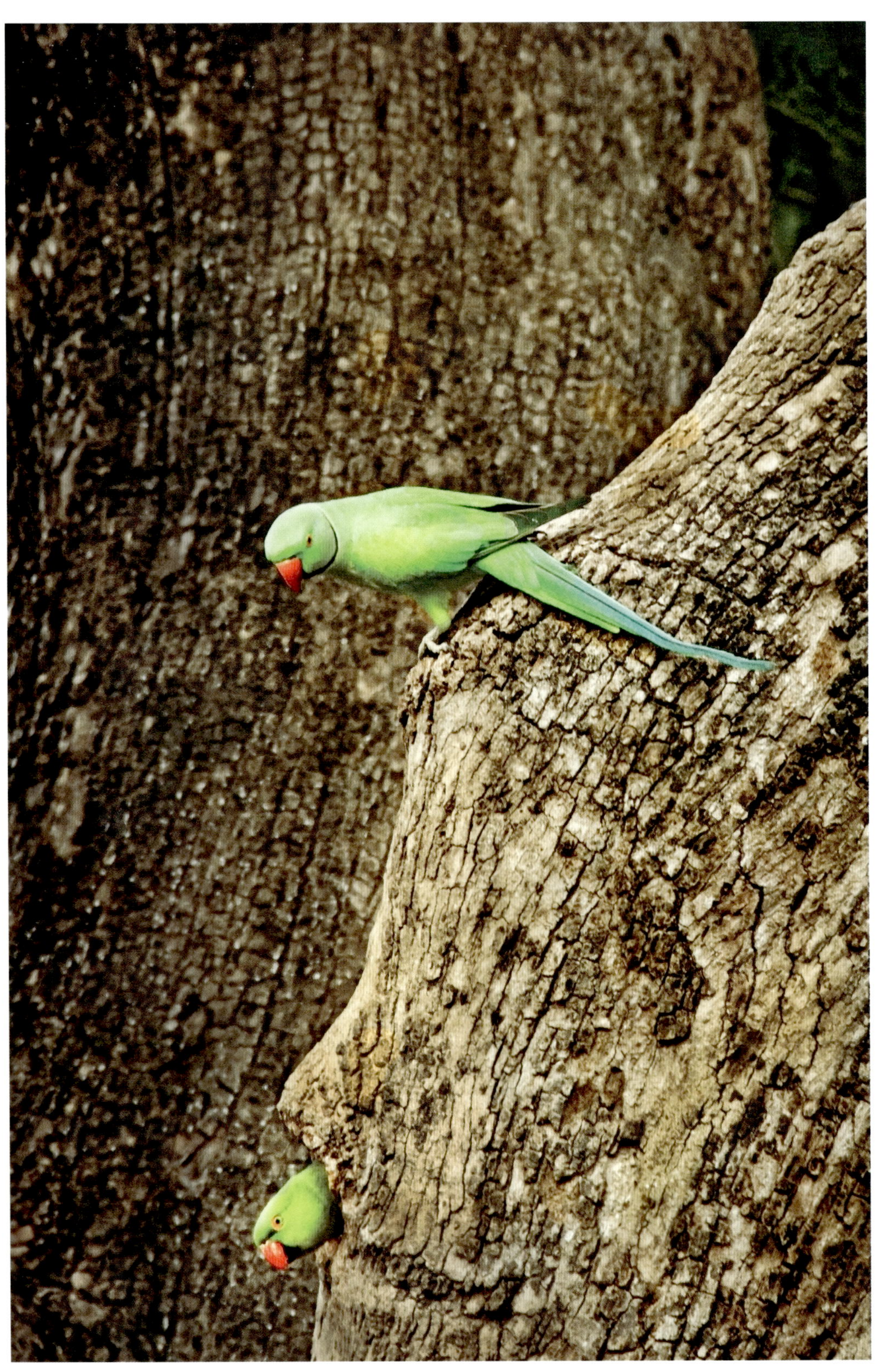

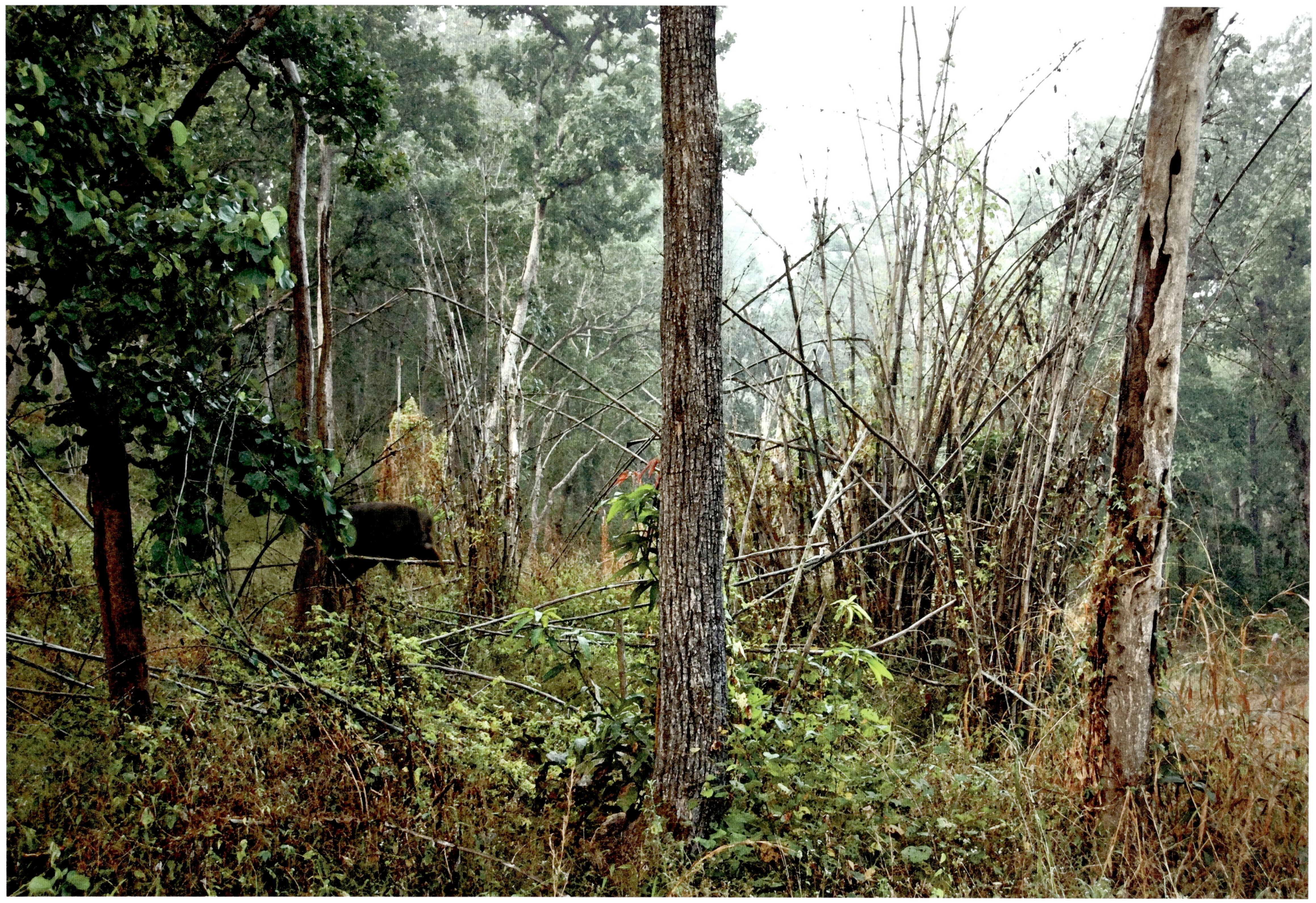

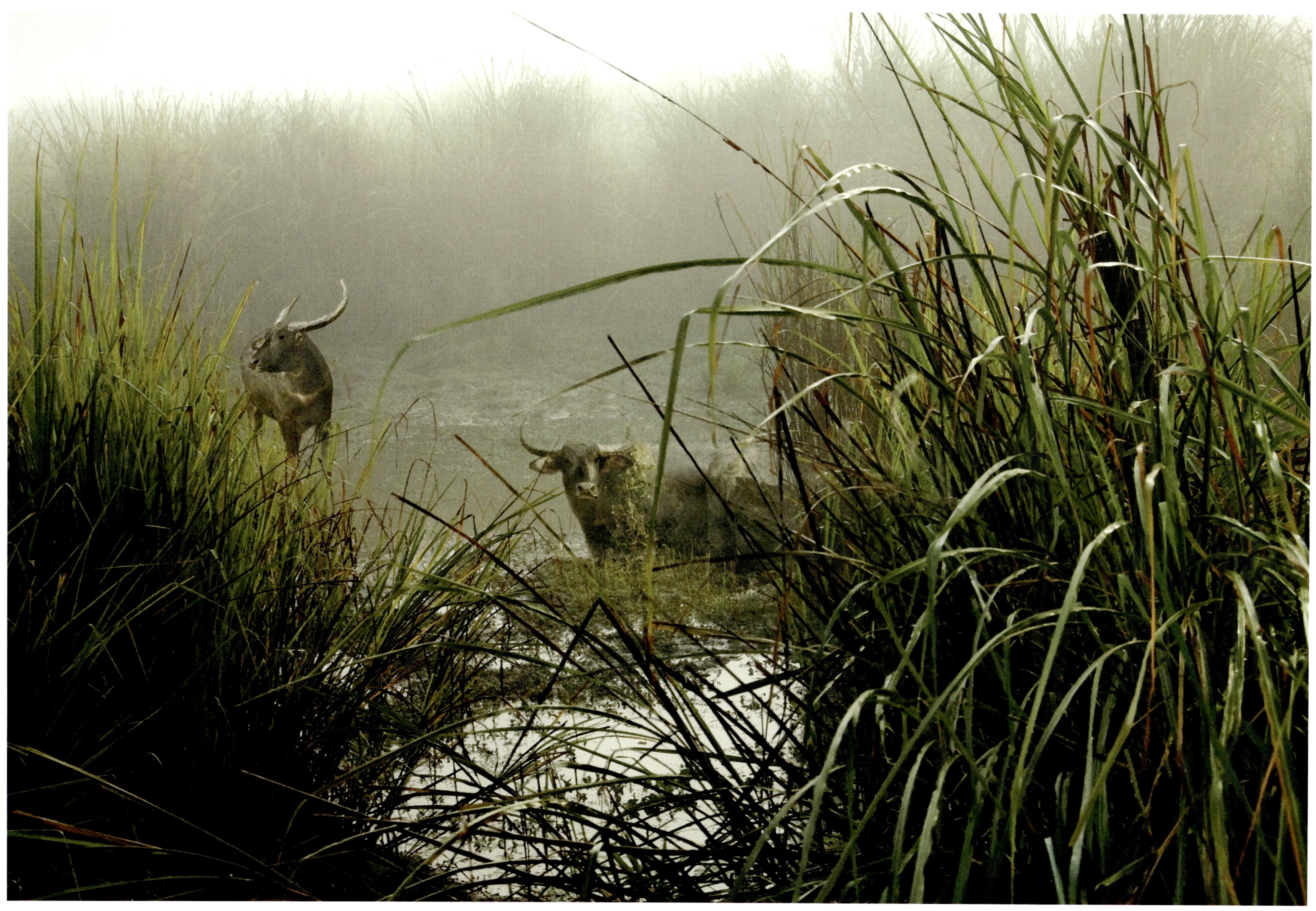

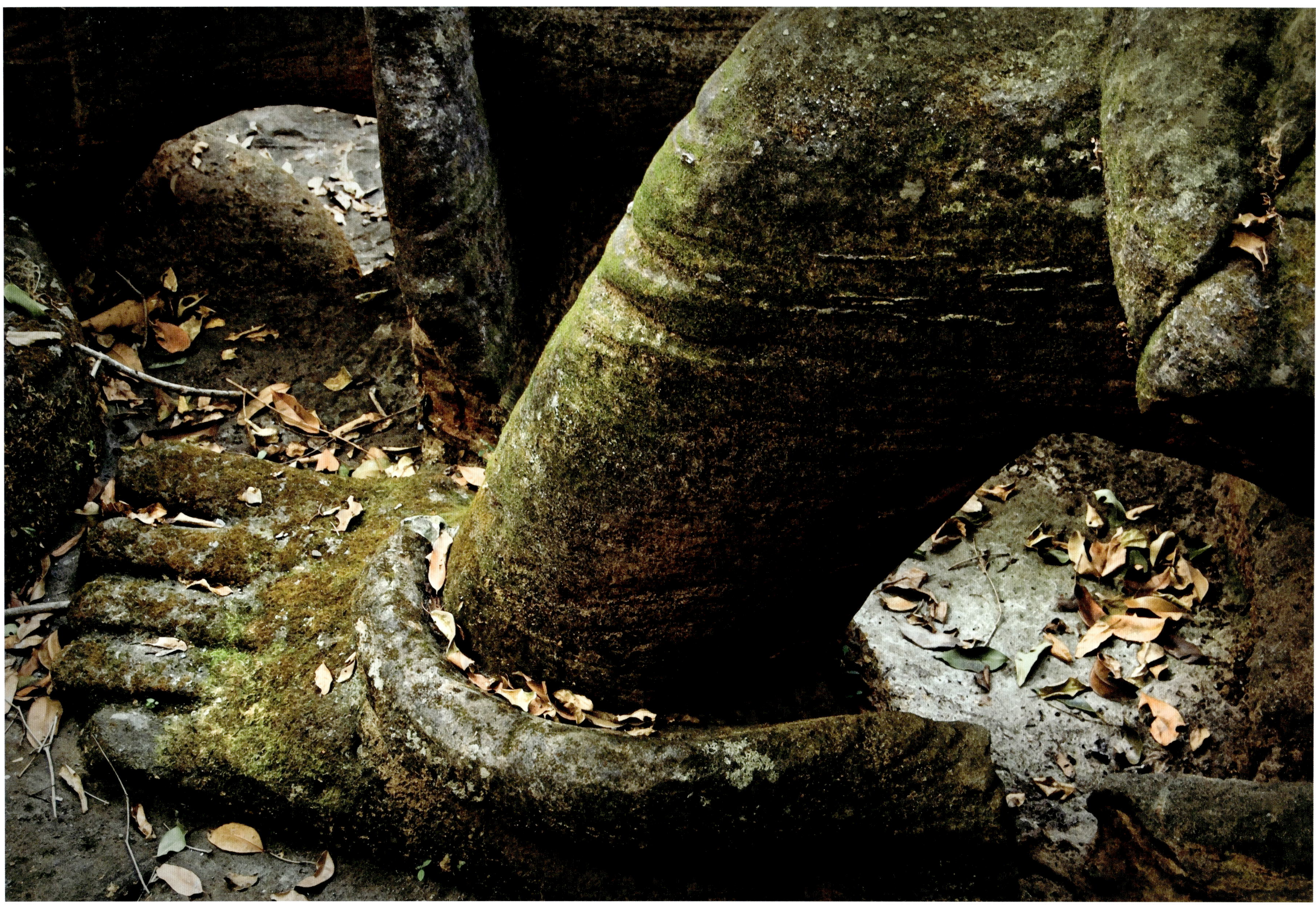

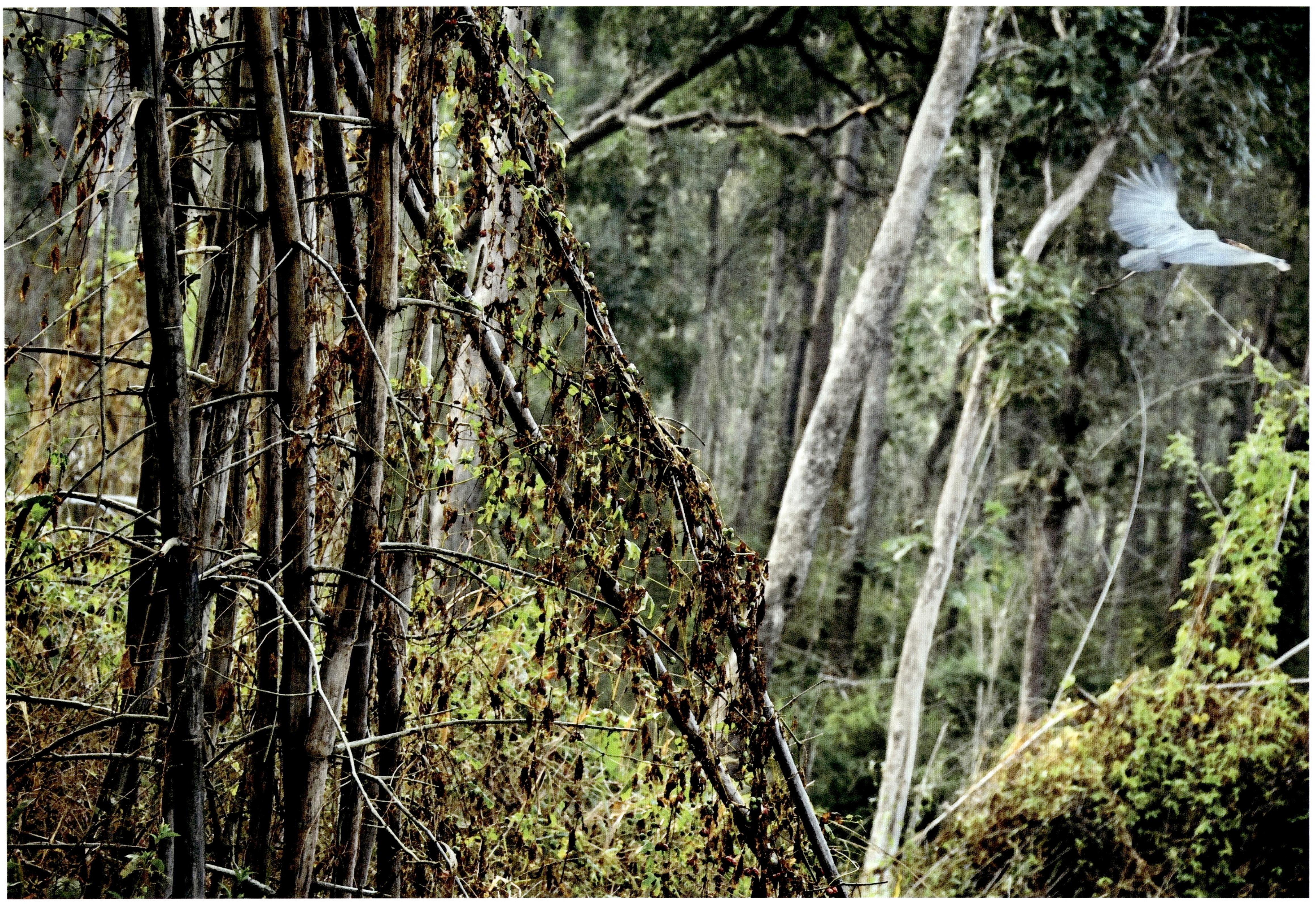

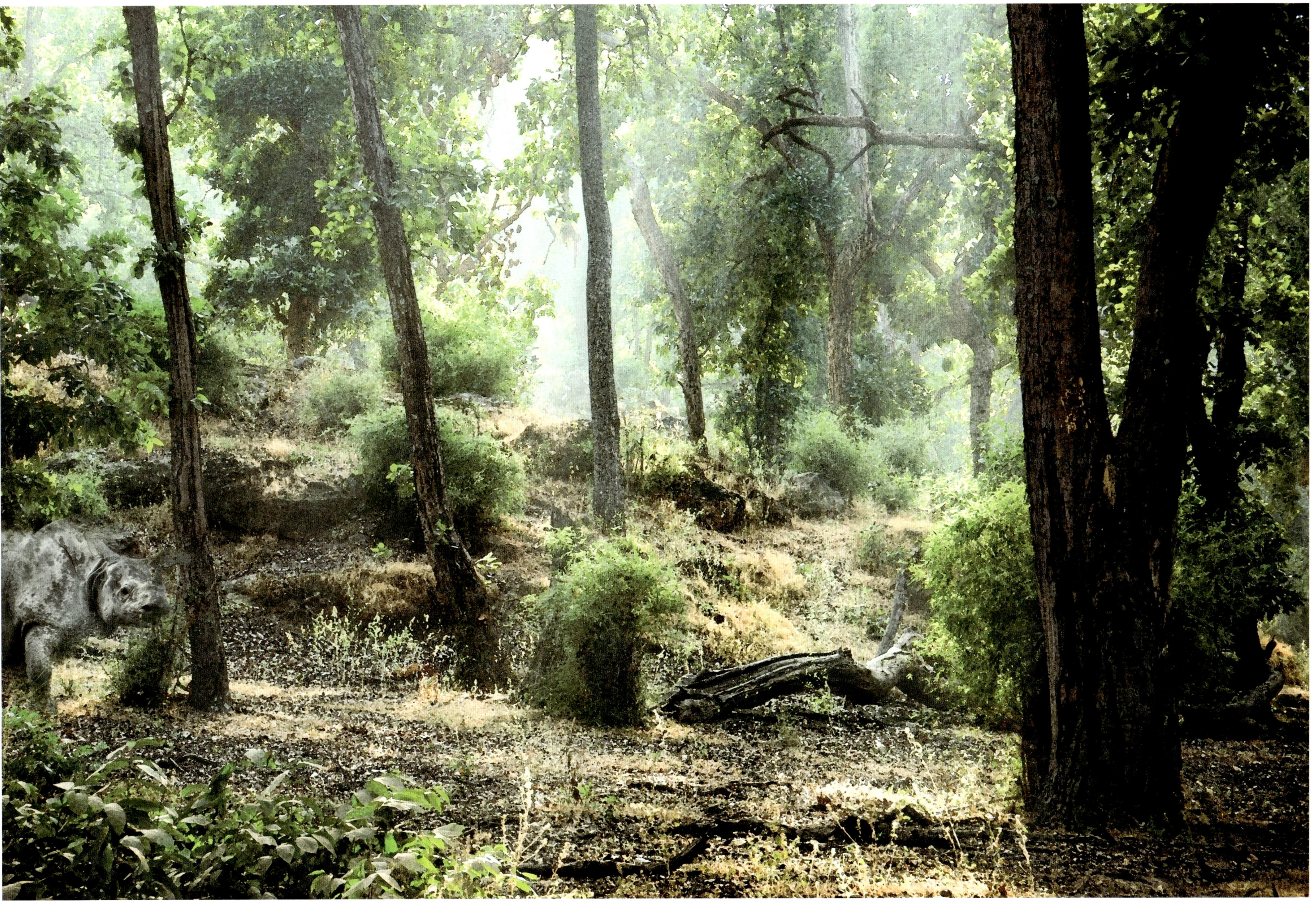

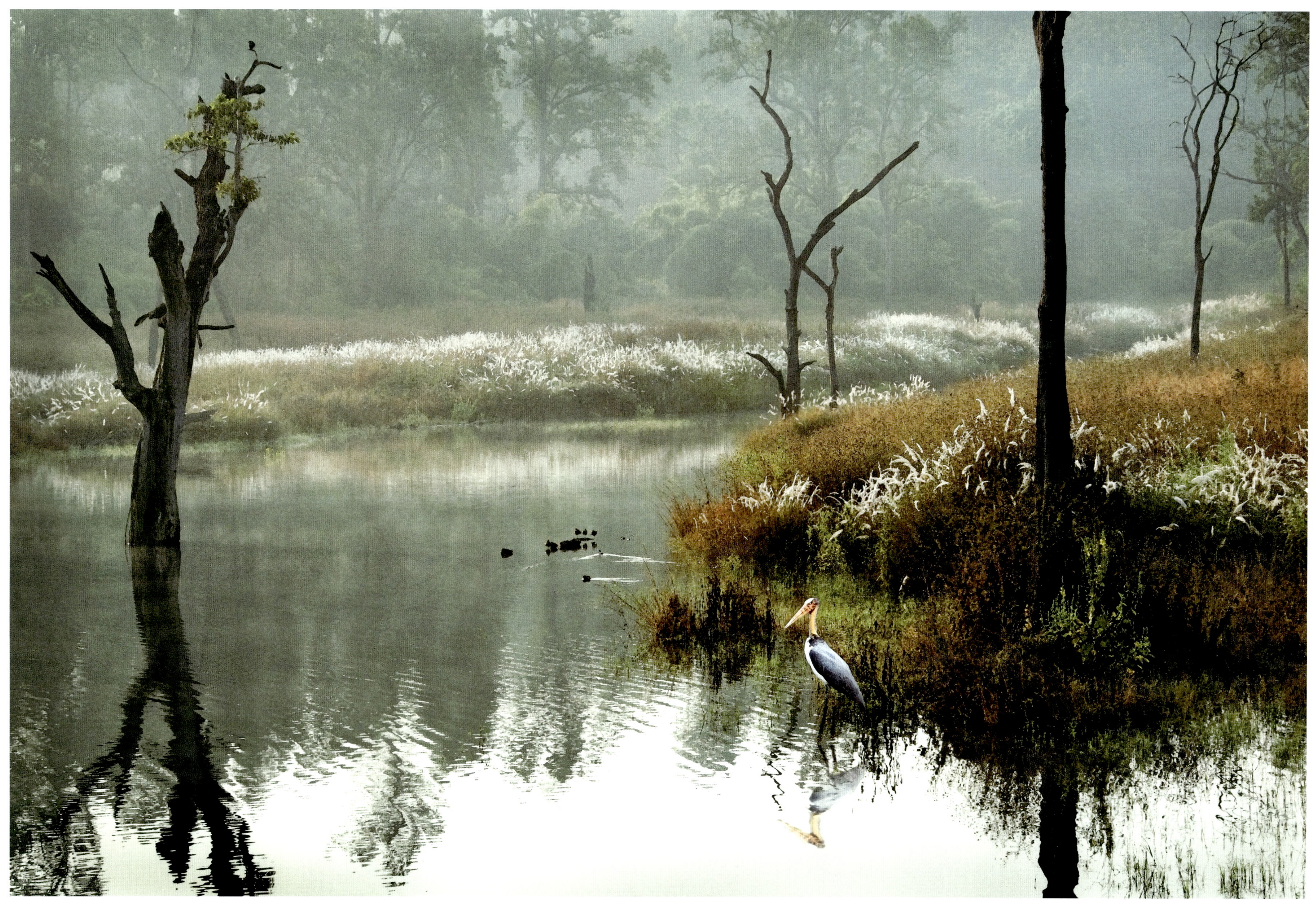

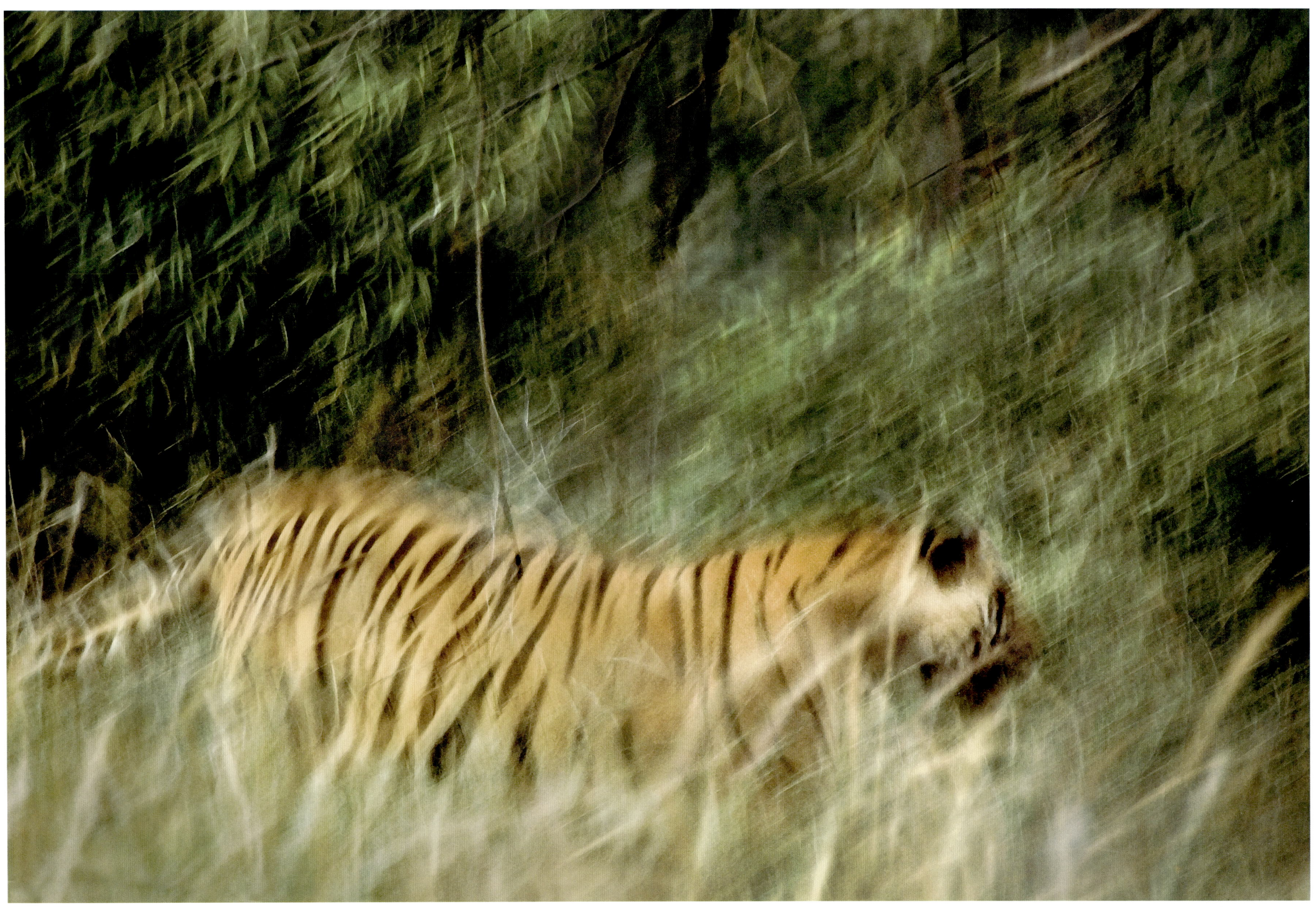

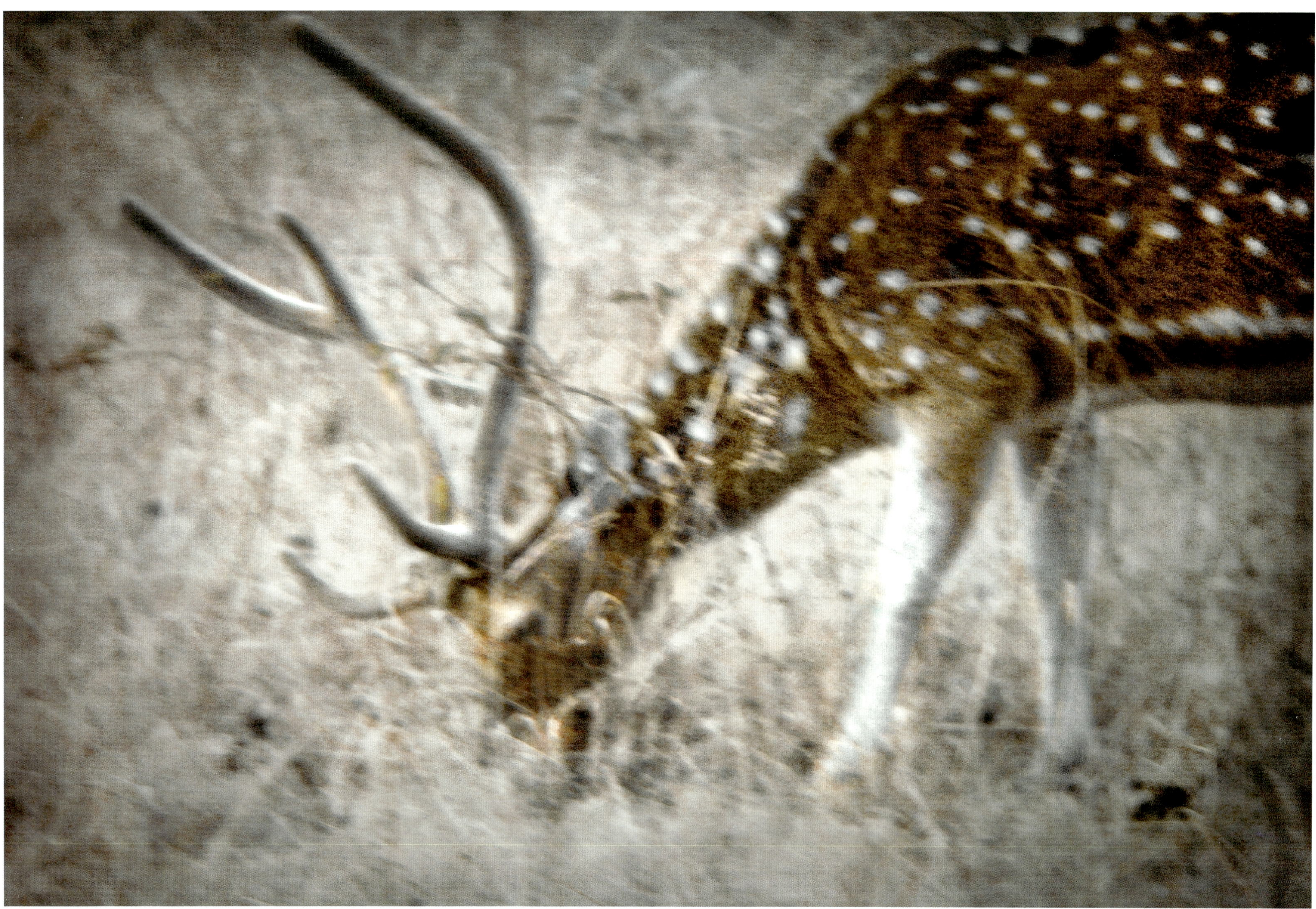

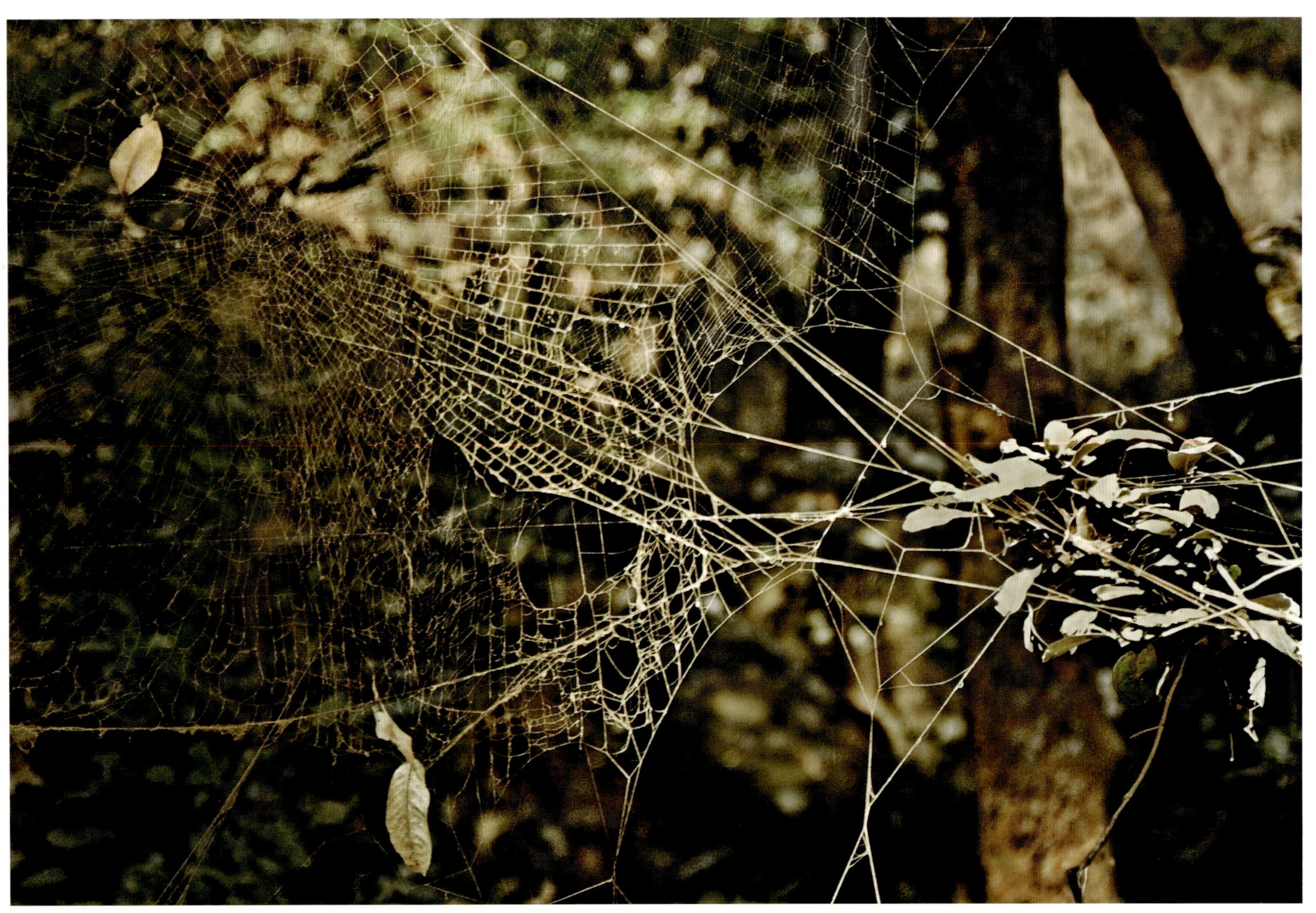

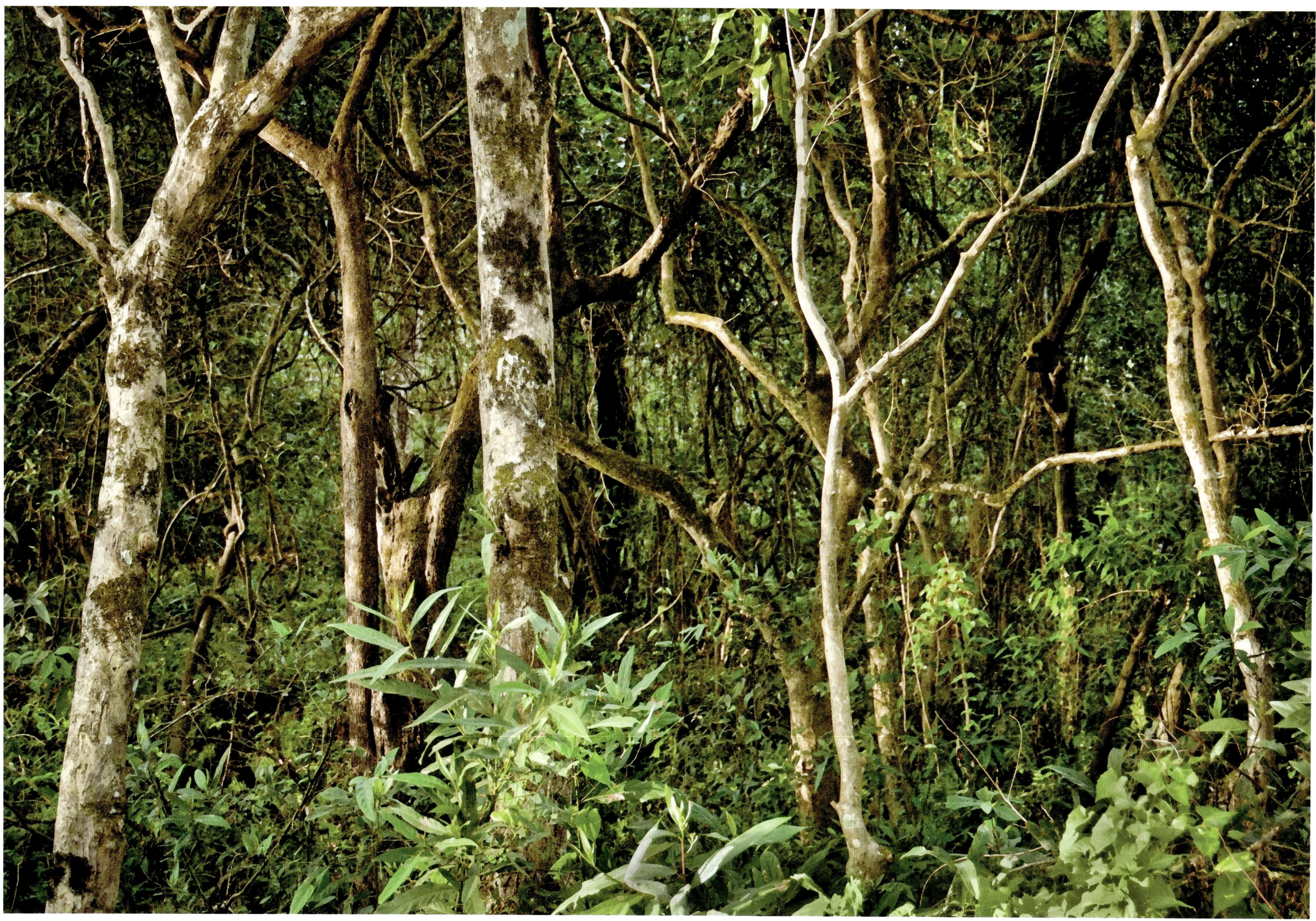

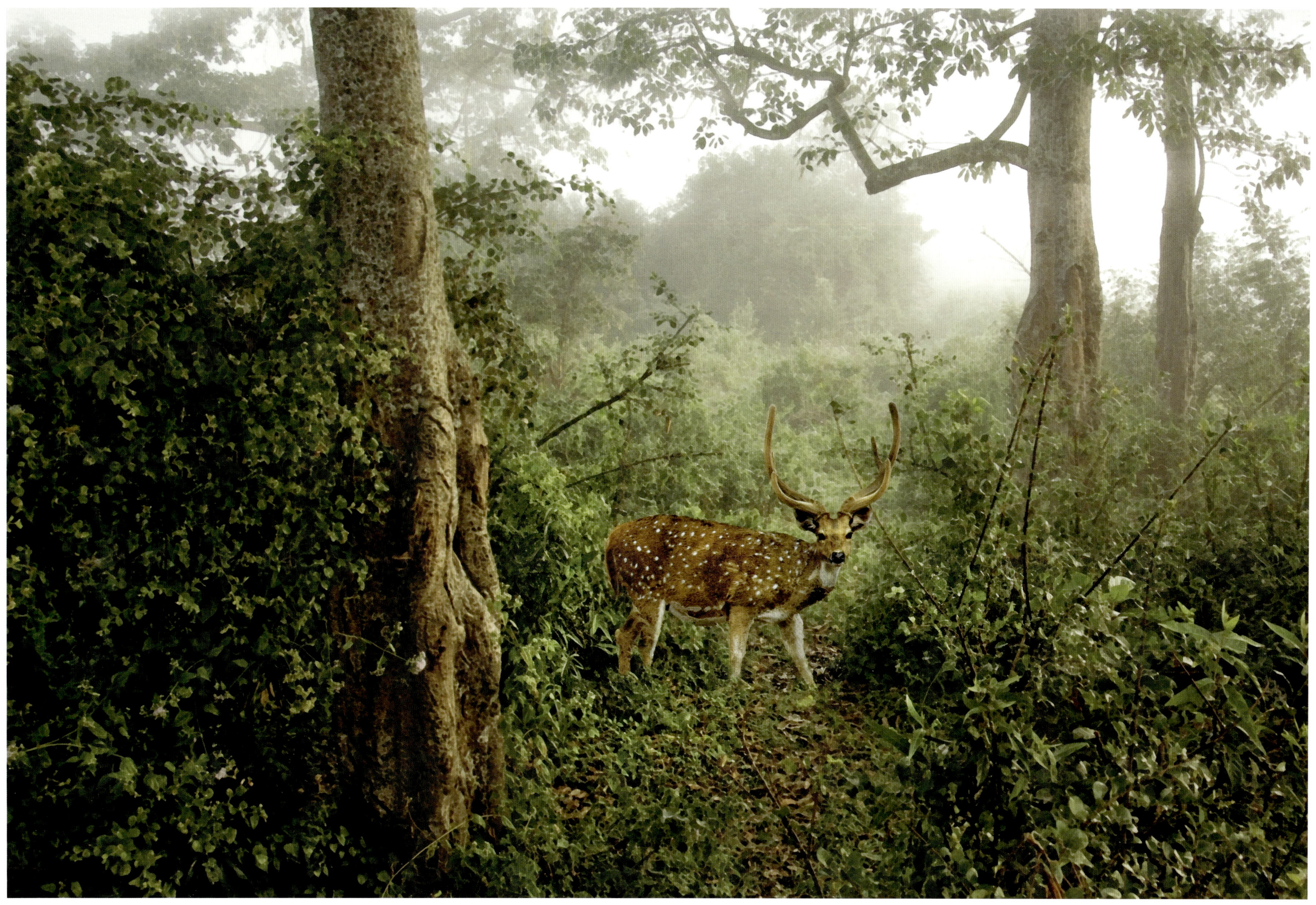

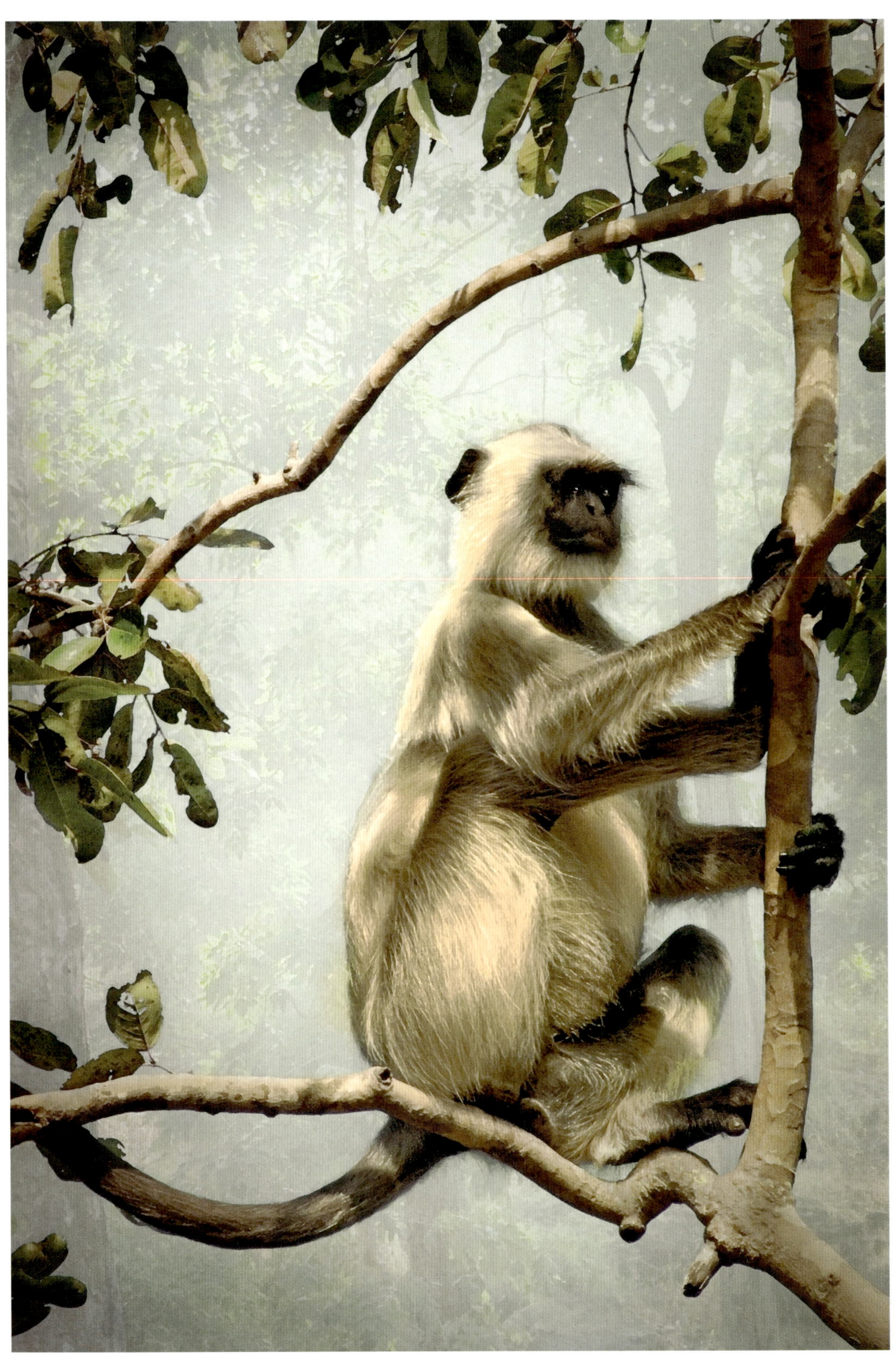

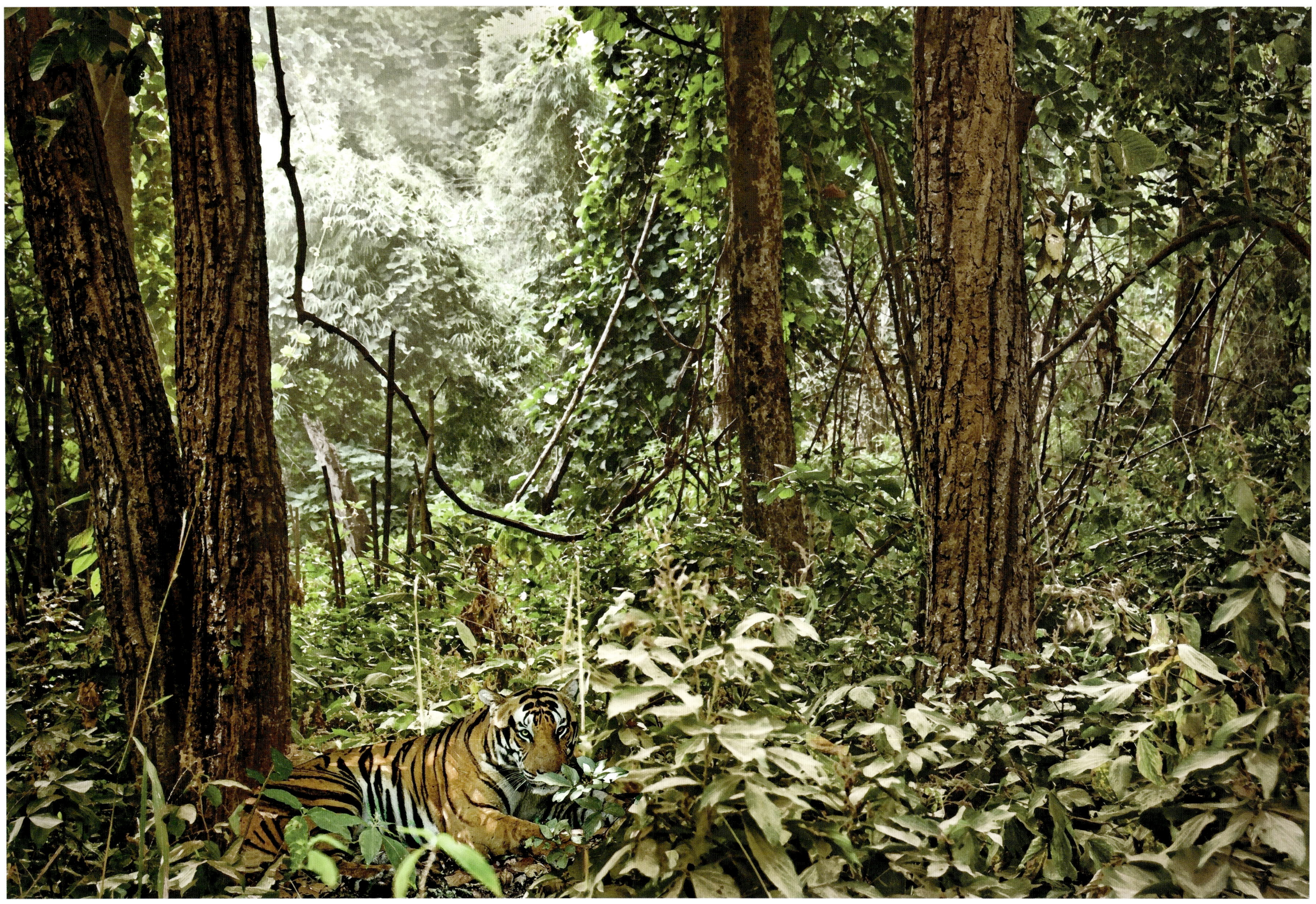

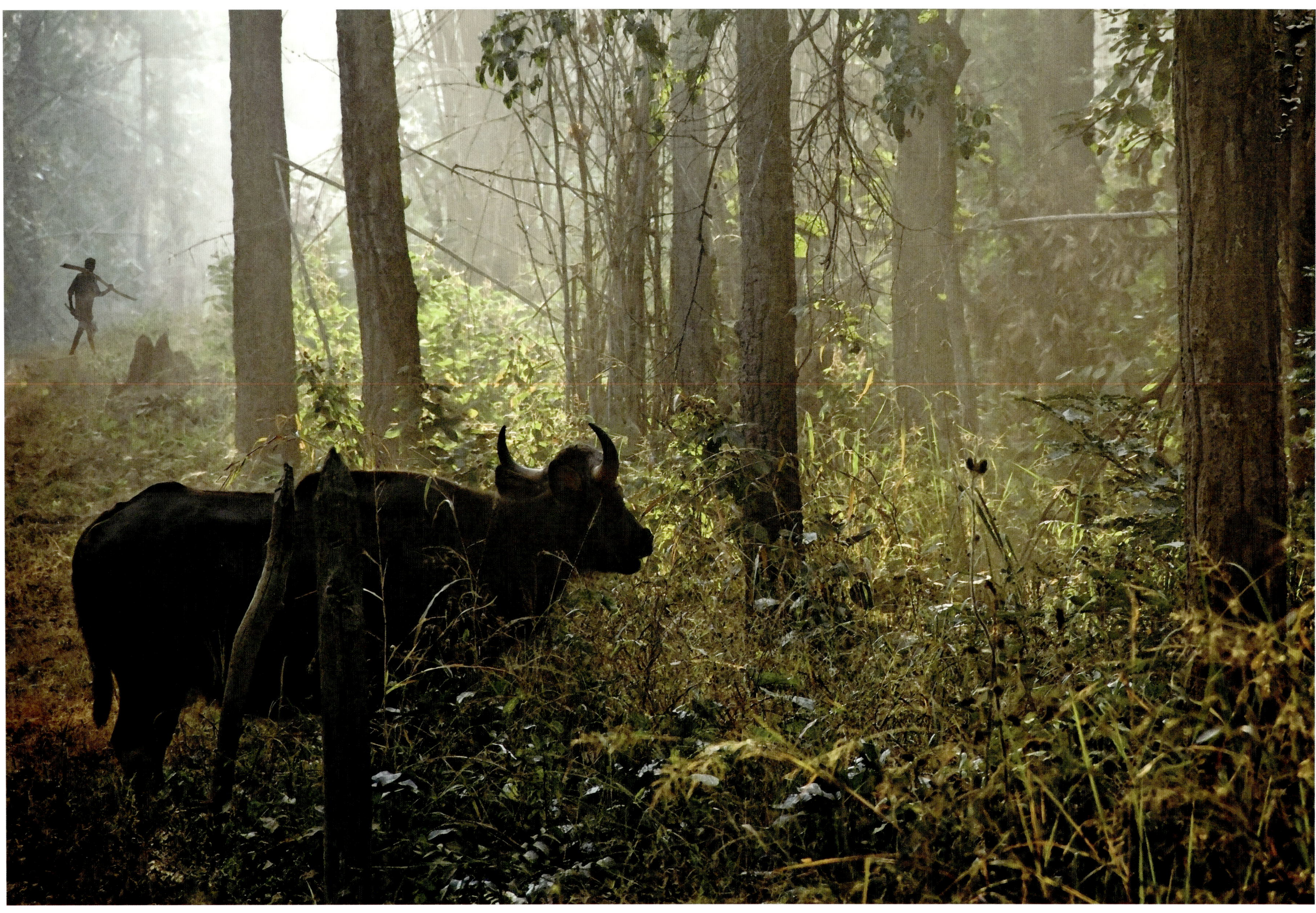

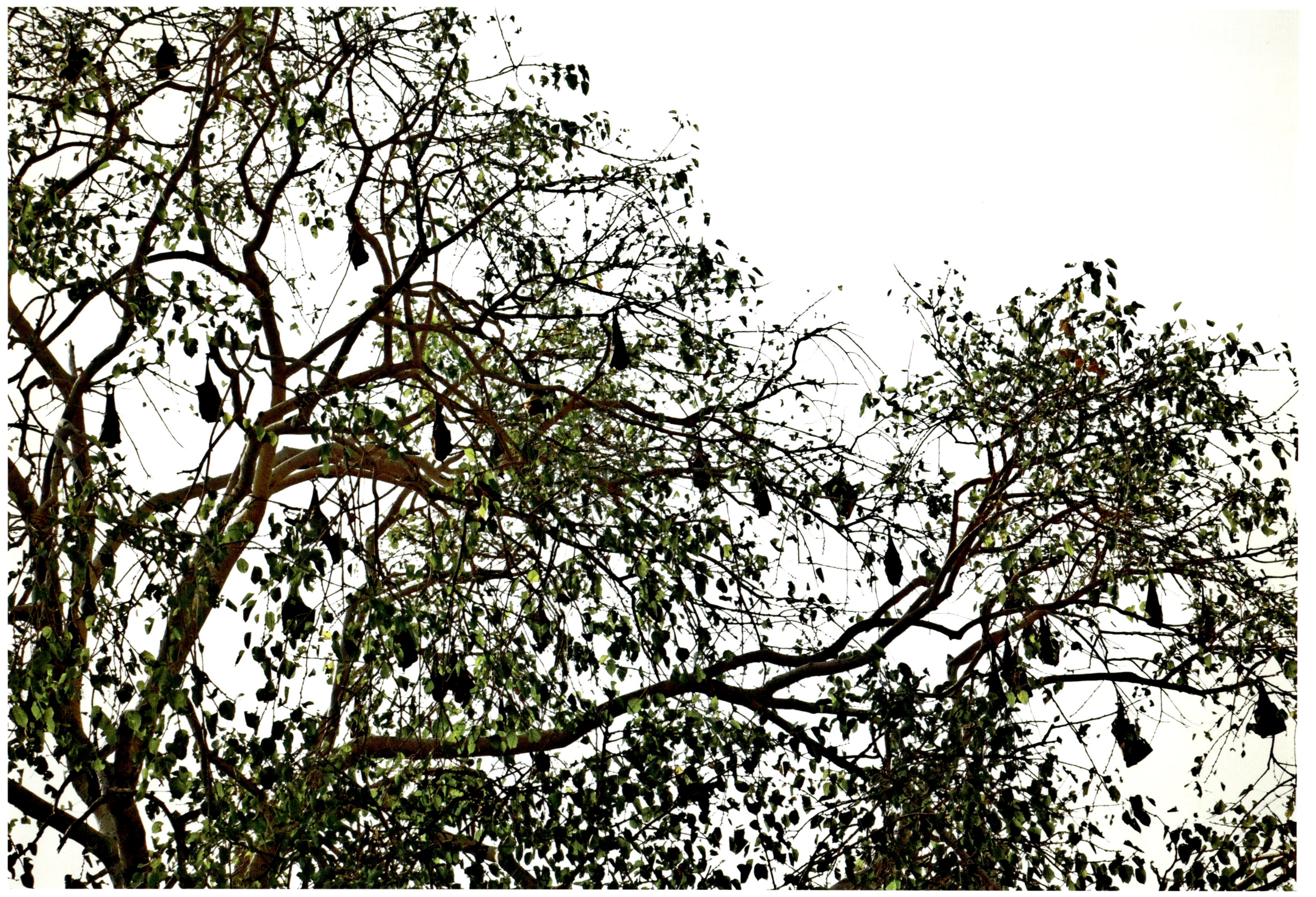

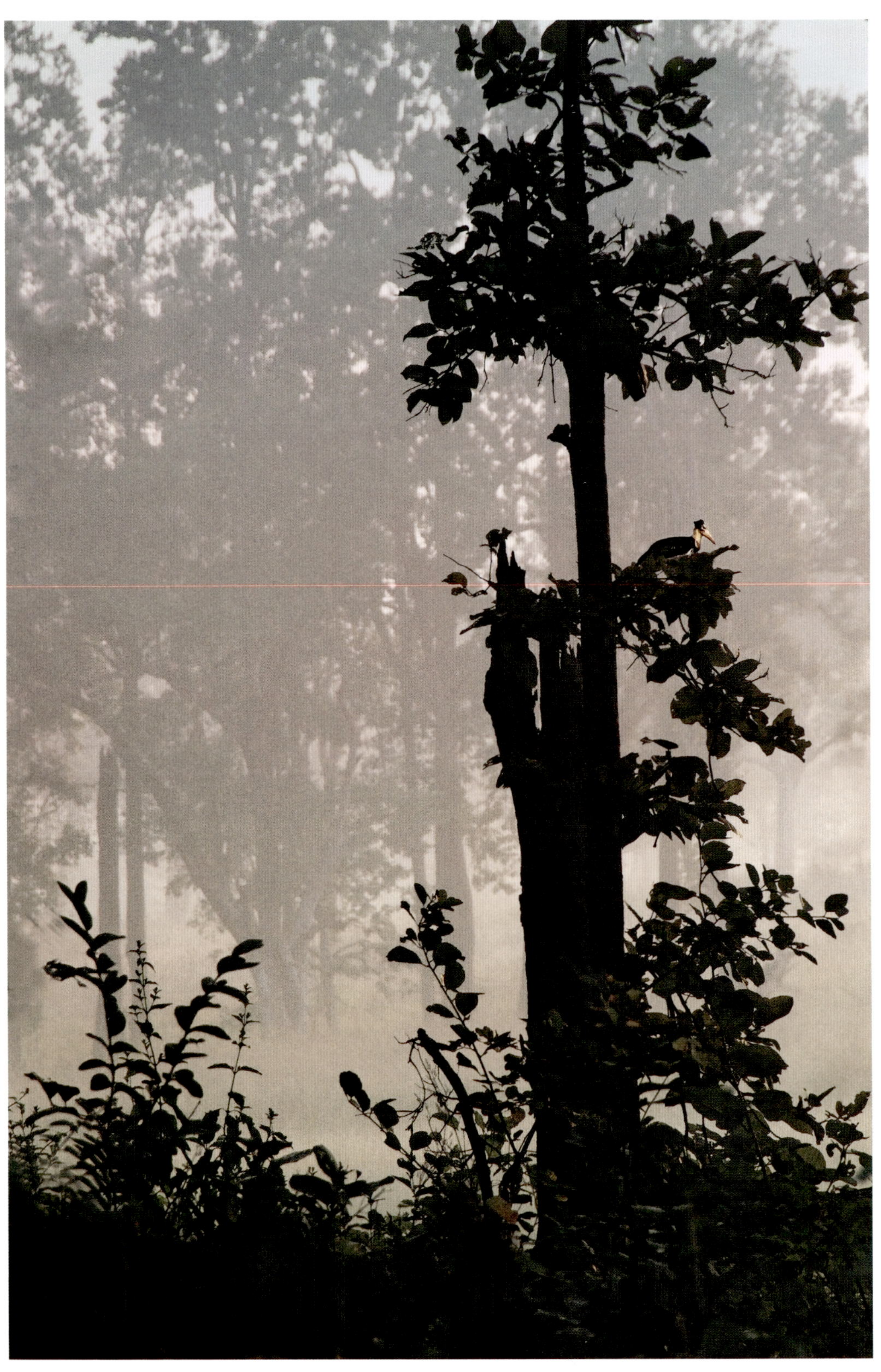

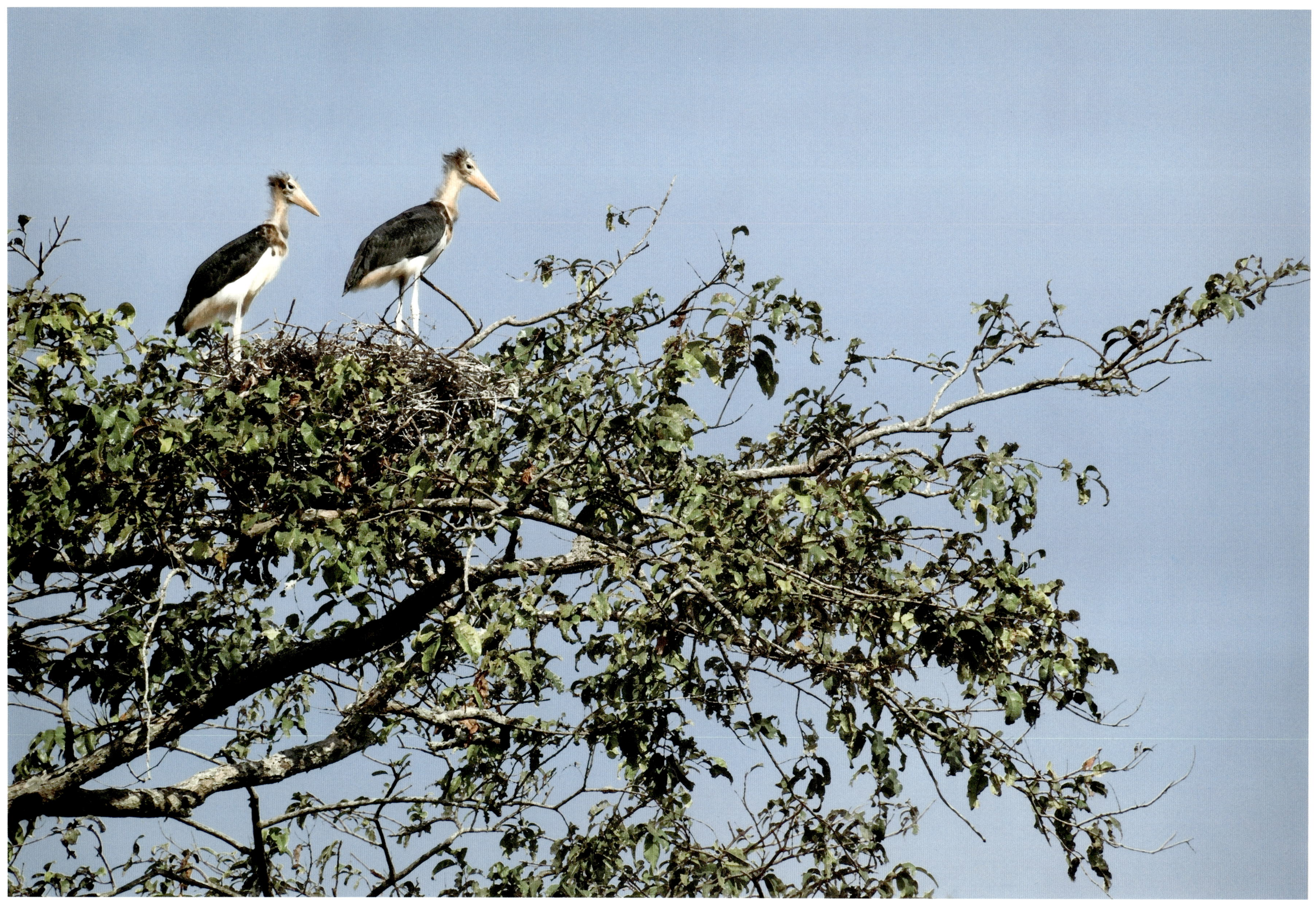

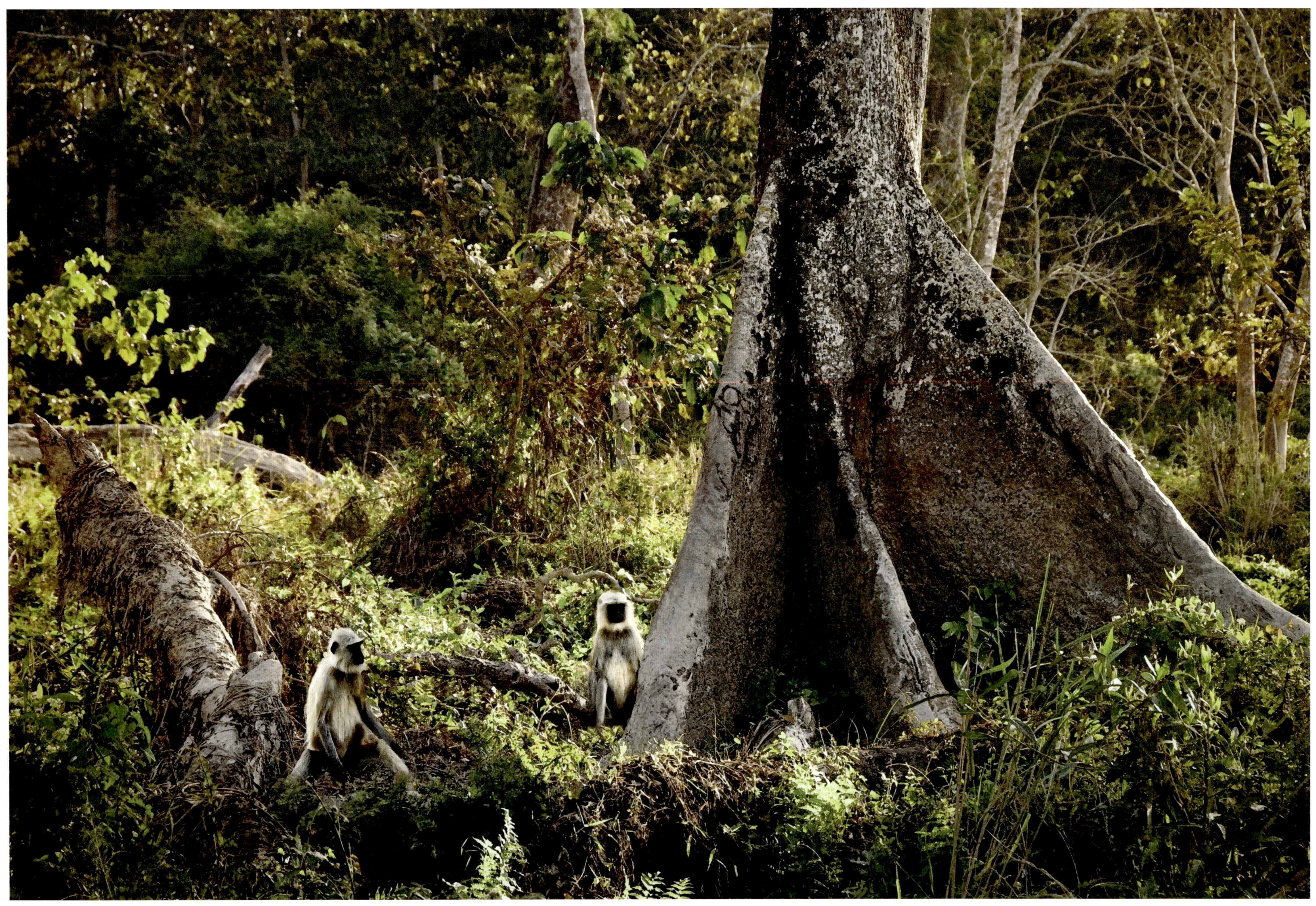

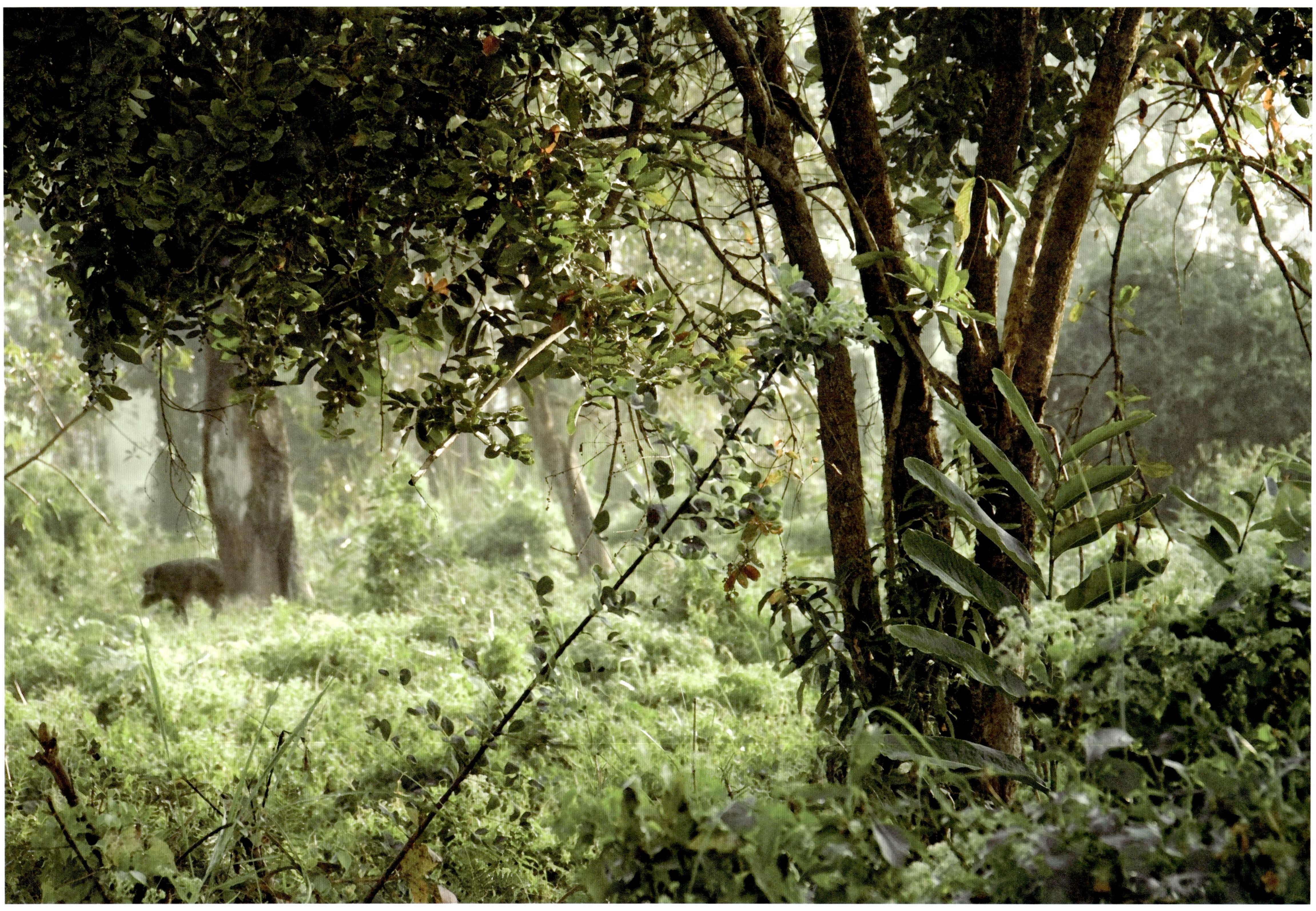

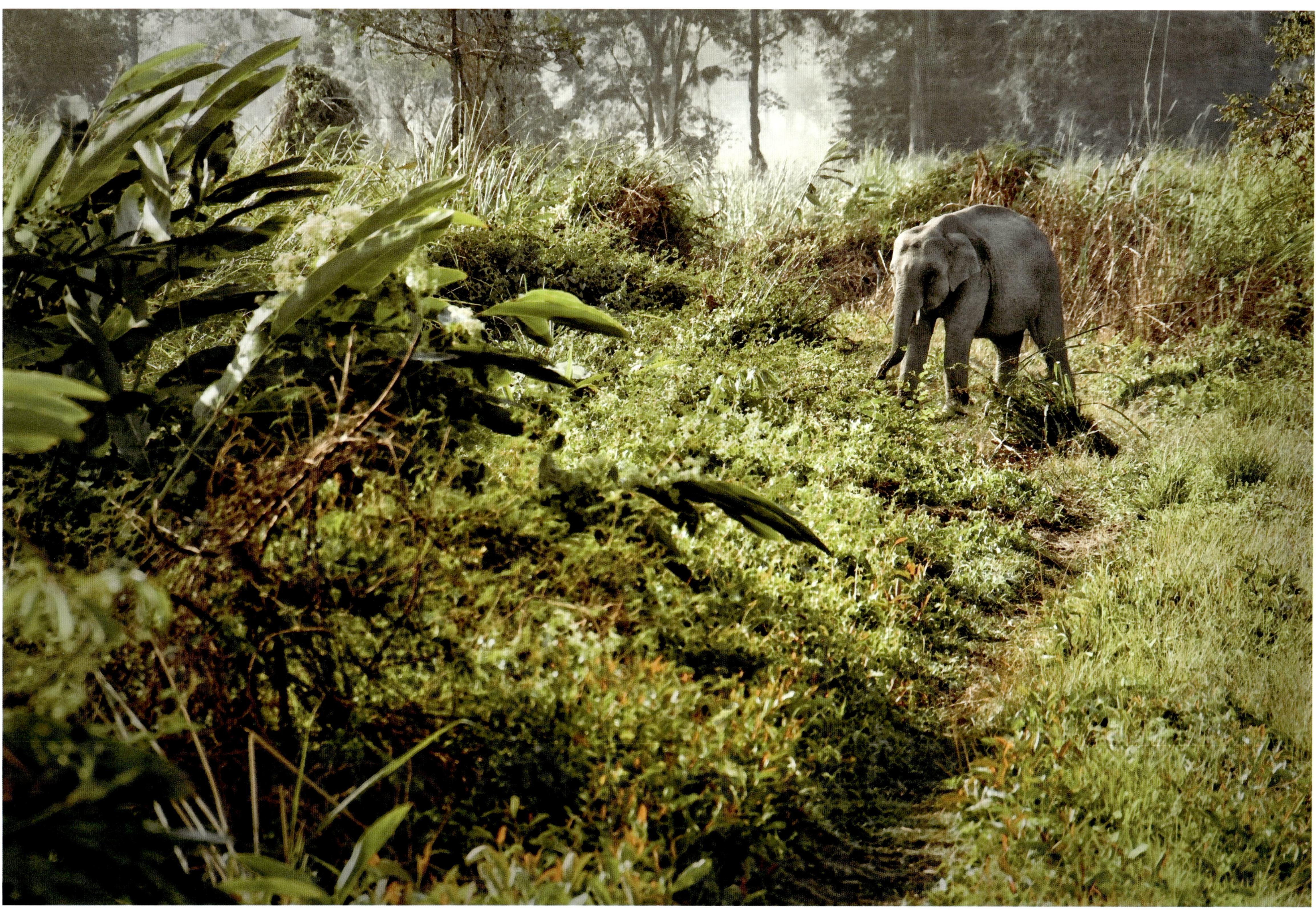

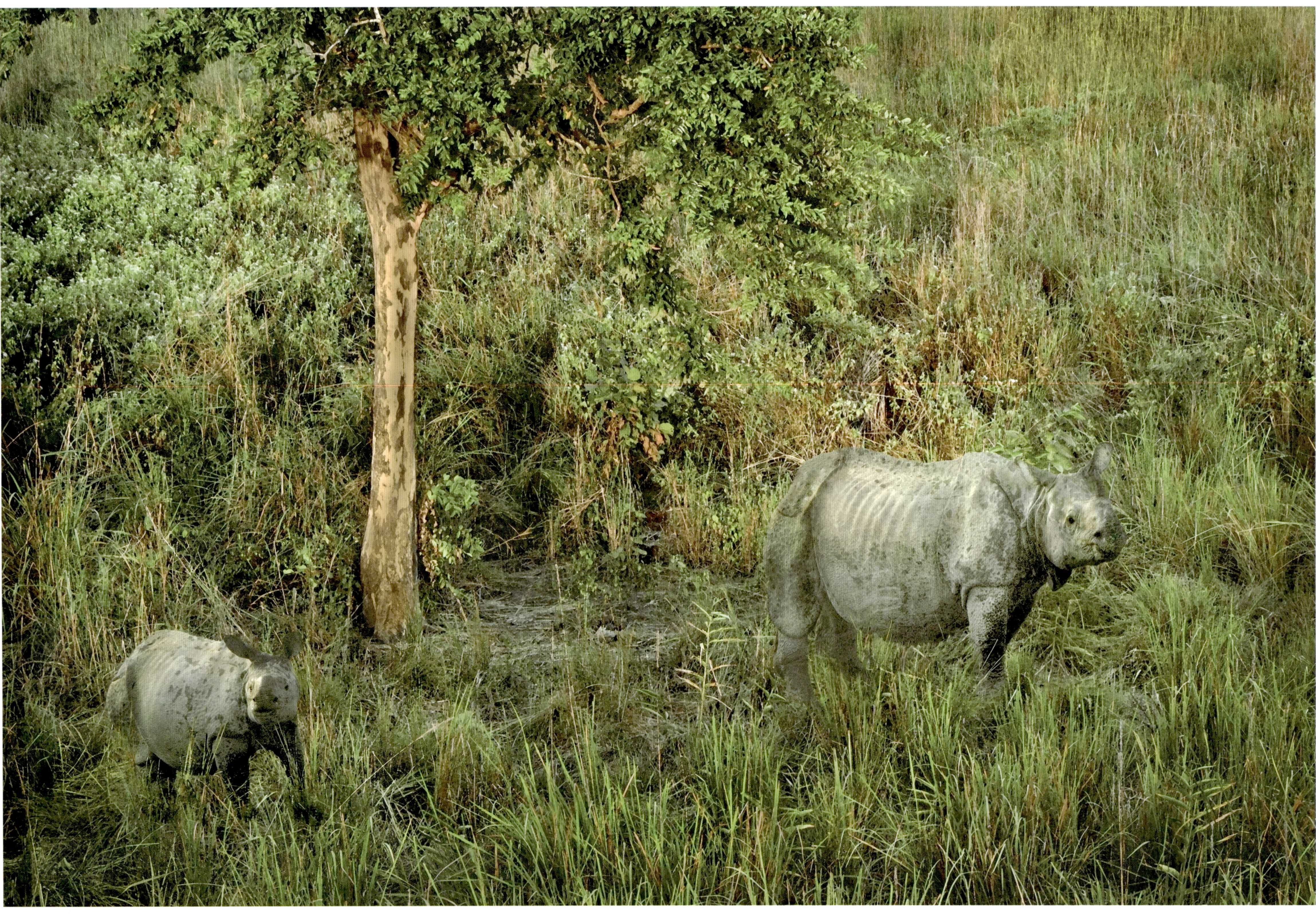

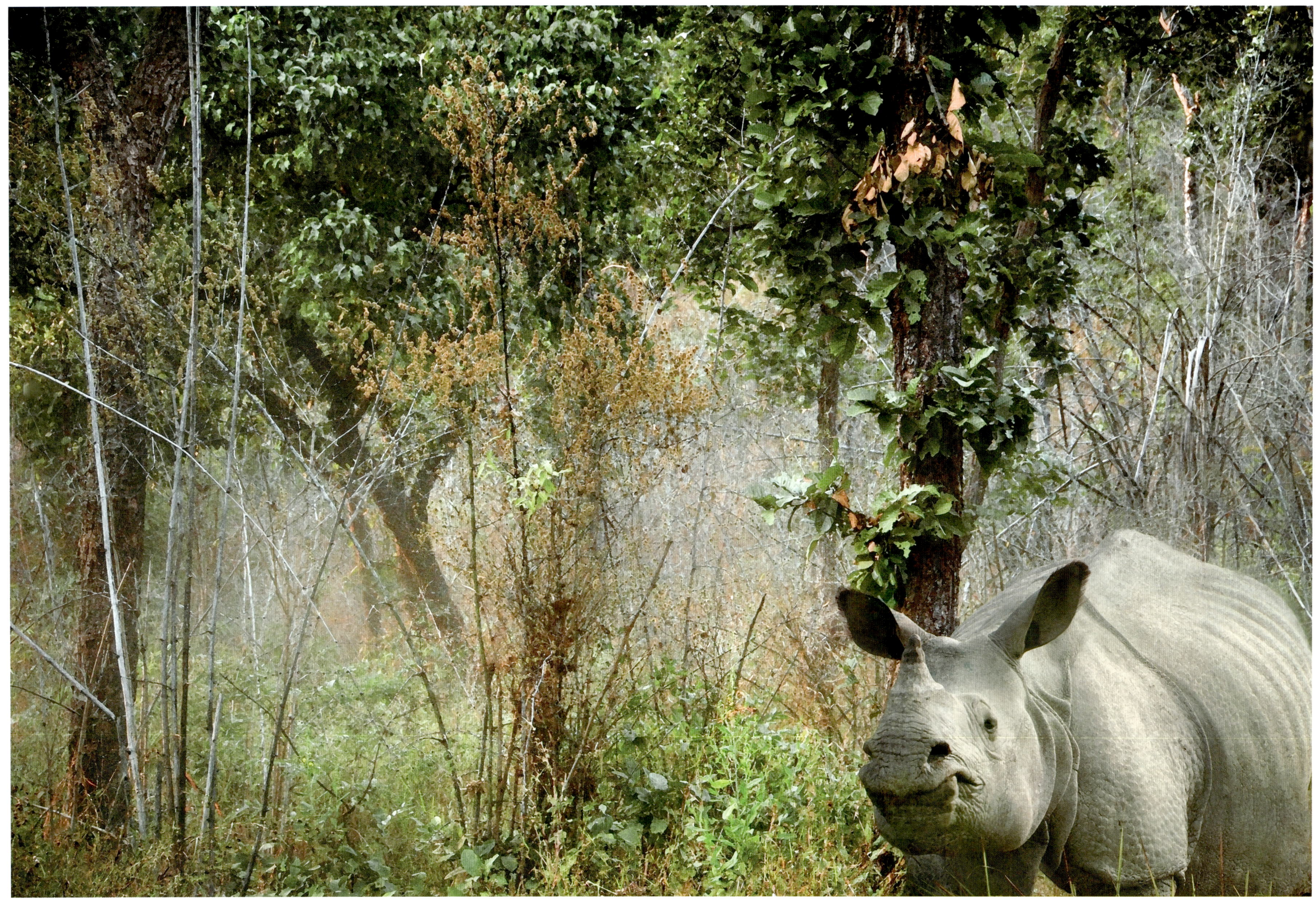

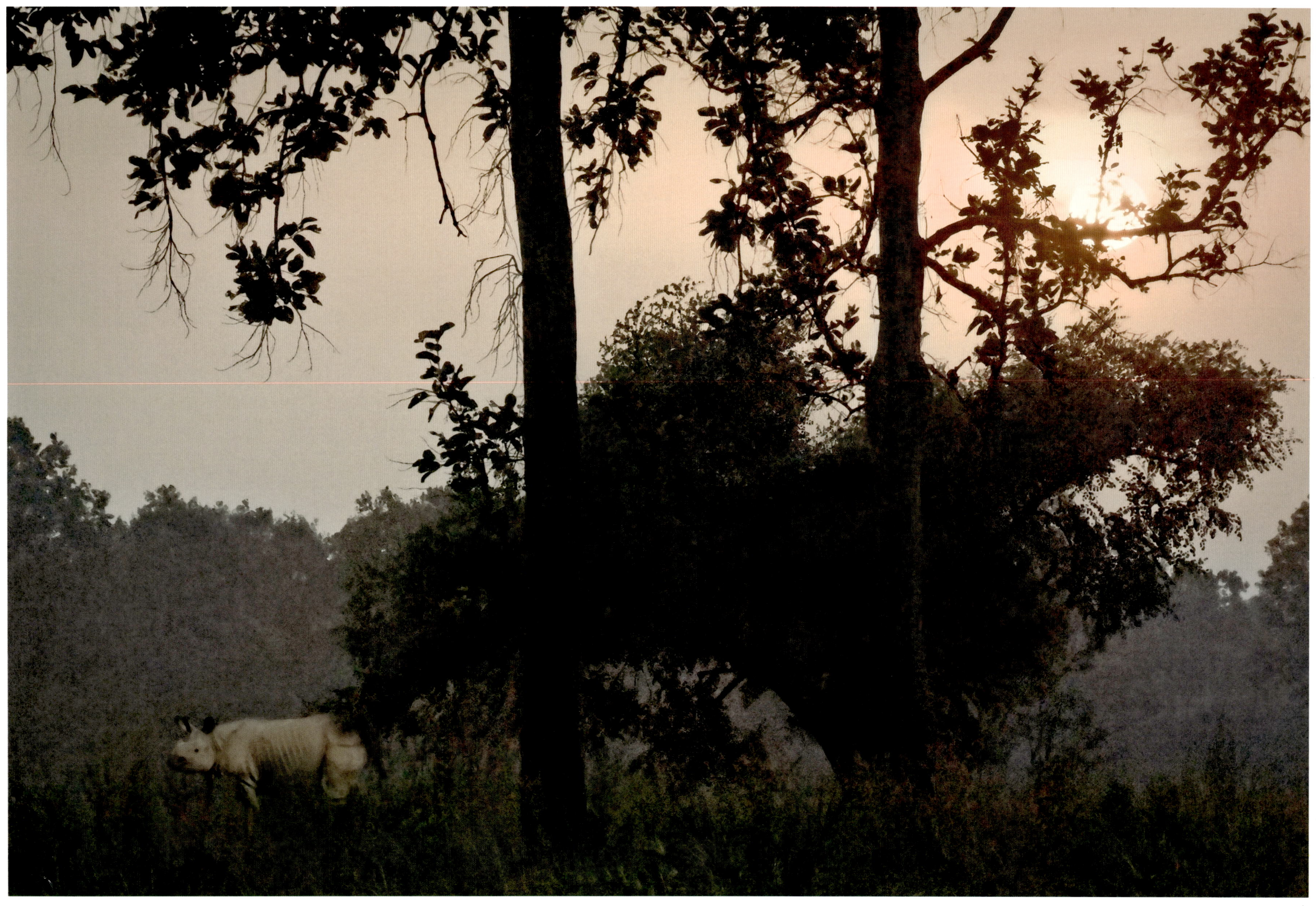

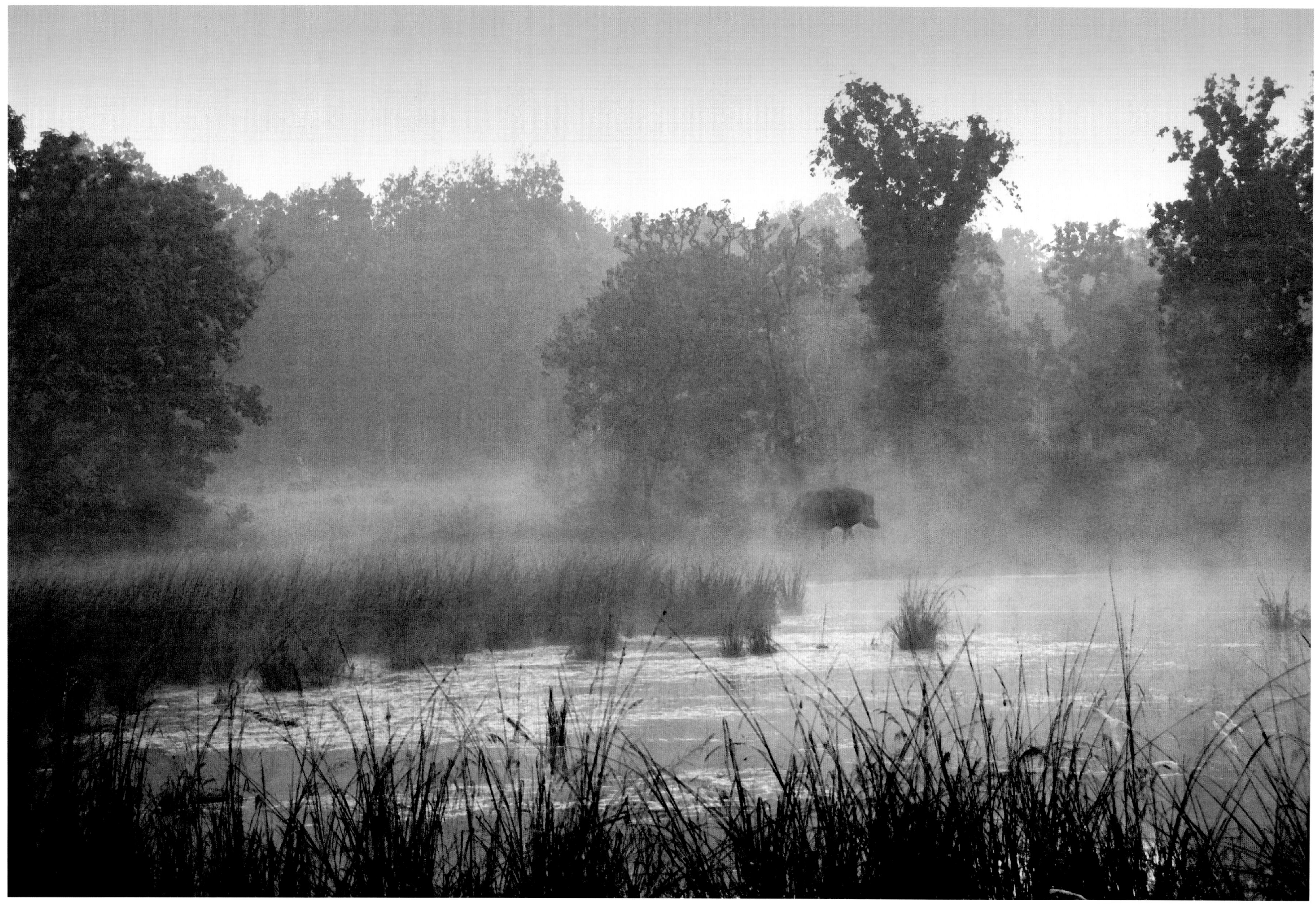

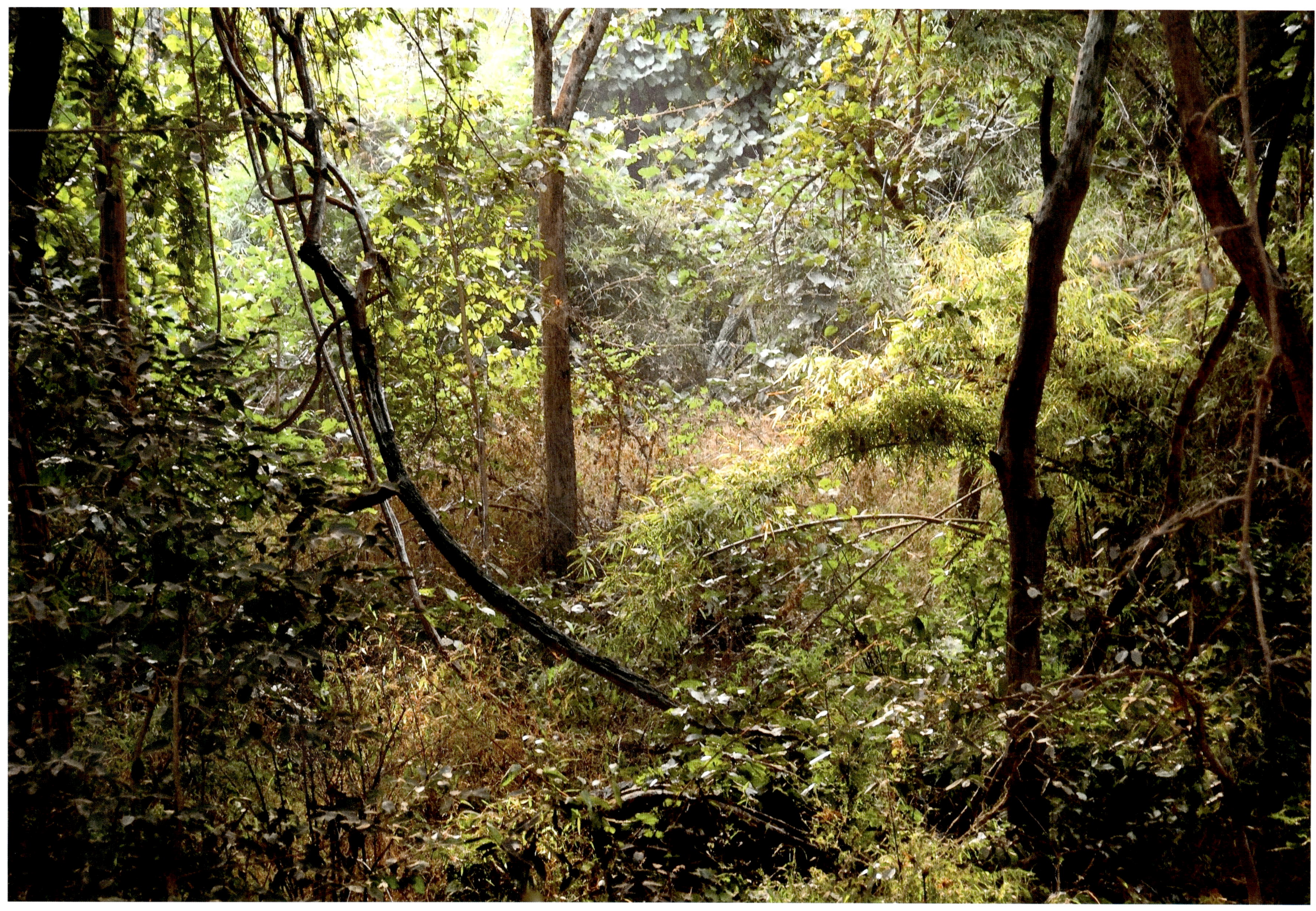

Jungle Journey

Joan Myers

Jungle originates from the Sanskrit word *jangala* and is prevalent in many languages of the Indian subcontinent, particularly Urdu and Hindi. Originally, it meant "arid, sparsely grown with trees," and, later, "land overgrown by vegetation in a wild, tangled mass."

Unlike many of my friends, I was never tempted to visit India. It seemed too crowded and busy, uncomfortably humid, and a little dangerous, especially for someone carrying a camera. Its cities thrive with a culture both unfamiliar and intimidating. The religious sites in India are spectacular but floridly ornate. My loves have always been dry, harsh, and barren places like the deserts of America's Southwest, the fringy places where people struggle to live in the empty space between land and sky. India's landscape, far from minimal, is fully and extravagantly occupied.

Still, as a small child I loved Kipling's *Just so Stories* and *The Jungle Book*. I would turn the pages in my illustrated edition, even before I could read the words, looking at the drawings of the boy who lived with wolves and played with a panther and a bear. I loved the monkeys with their mischief-making. Shere Khan, the tiger, both attracted and frightened me. I practiced saying "rikki-tikki-tavi" over and over and imagined being there in the jungle to see the battle of mongoose and cobras. Kipling's India was a magical, mysterious place.

So, when I had an opportunity to visit wildlife refuges in central and northeastern India with a small safari of bird-watchers, I made the long flight with trepidation and curiosity. I spent several days in Delhi, where I was swept away by the bright colors and the smells, noise, and seemingly limitless variety of human activities visible on the streets. By the time I departed the city and took another flight south, followed by a lengthy drive, I felt dazed.

I had forgotten all about Kipling and the enticing forests until I found myself seated in the back of a four-wheel open van entering Bandhavgarh National Park in central India. It was 5:30 a.m. in late November and so cold I had on many layers of clothing, a hot water bottle on my lap, a wool blanket wrapped around me and still shivered in the chill before sunrise. My teeth chattered.

"Listen," our enthusiastic young guide whispered. "We might see a tiger…Listen, you can tell from the warning calls of the monkeys and the deer."

The light was low. It was too dark, I thought, to see anything, much less to photograph. Yet, the tension of driver and guide was palpable. I shivered and listened. In half an hour or so, a hazy sun splashed gentle light into the forest. It was still cold and dim, but I uncapped my lens and continued to listen.

"Look," the guide said. "A tiger track in the sand. Fresh. Hear the monkeys calling? The tiger is nearby and hunting."

The surprise came when I put the camera to my eye. Instead of confused underbrush, vines, and trees, I saw Kipling's jungle. Unfamiliar trees, vines, and ferns flourished. The jeep stopped frequently for the guide to point out another racket-tailed drongo, one of some 300 birds and thirty species of mammals eagerly spotted and identified by my enthusiastic birding companions.

Bandhavgarh National Park was formerly a maharajah's private hunting preserve that was declared a national park in 1968. It is famous for its tigers as well as barking deer, langurs, rhesus macaques, sambar, and a multitude of birds: bee-eaters, Eurasian golden orioles, hornbills, rufous treepies, and spotted owlets, most of them as spectacular as their names imply. Since I am not a birder, and many of these birds could be identified only through a spotting scope, whenever the jeep stopped I took photographs of the forest. In the early morning light, the vegetation seemed to glow gently, as if lit by a Hollywood lighting expert from a 1930s movie. Sometimes I was lucky when a wild elephant, or monkeys, or even a tiger passed through. In some areas, jungle gave way to open grassland, and I saw deer, peacocks, and rhinos.

At 8:00 a.m. that first morning, we ate a breakfast of hardboiled eggs, cold *pakoras*, cheese sandwiches, and tea. Gradually, the sun rose high enough to warm us. When I stopped shivering, I began to imagine all the animals hidden by the vegetation. Though I didn't fancy being a meal for a tiger, I longed for a walk in the forest. This was forbidden; the refuge's rules required us to remain in the van. This felt like being transported to Jurassic Park, knowing the danger but being unwilling to remain secure and not explore what lurked in the underbrush, just out of sight.

Our first tiger-sighting came early in the morning several days later, when a large male emerged from the jungle like a creature from a waking dream. His head was enormous. His skin rippled as he walked along. He seemed to float above the ground. What an elegant creature a tiger is! He came out of the bushes lining the dirt road, peered out from behind a tree, opened his mouth, and growled in a rumble that seemed to shake the forest. He paid no attention to our van, but I could feel my hair stand on end.

It is hard to see a tiger, even if it is large and gorgeous. Trees, vines, and underbrush clog the forest, and clearings are dense with high elephant grass. Tigers prowl when they are hungry, mainly in the early morning when the light is dim; they then find a comfortable resting place in the underbrush, where they are well camouflaged for the warm daylight hours. Early in the morning, then, is the best time to see one. They have incredible vision, their sight enhanced by a reflecting layer behind the retina that enables them to see six times better than humans in dim light. As a less visually endowed creature, I knew I had to set my camera speed high enough to photograph in the murky pre-dawn light.

Another morning when tigers weren't forthcoming, our guide took us on a short excursion to a statue of a reclining Vishnu, sculpted in one piece from the rock of the mountain in the center of the park. In front of this twenty-foot statue is a dark pool of spring water, also cut out of mountain rock. Ferns and flowery shrubs flourish in the moist surroundings. Due to the year-round availability of water at this pool, it is well frequented by animals, including tigers and monkeys.

To get from one national park to another, we often traveled in a van for hours through the Indian countryside, a journey that was taxing yet meditative. Sometimes it appeared that the roads had been recently widened and paved; other times they faded into dirt tracks, rough with boulders and ditches. Sometimes the roadside was filled with men, women, and children laboriously chipping away at large rocks until they had reduced them to heaps of tiny gravel for the roadway. Once, late at night, we reached an impasse at a rushing creek too deep to cross and had to backtrack several hours around the dead end. In the darkness, we drove through a small village. Houses and shops were mainly dark or lit minimally by candles or a single, bare fluorescent bulb. Everywhere villagers were shopping, carrying heavy loads home, cooking, eating, chatting, even getting their hair cut. The driver stopped suddenly. Ahead of the van a crowd spilled into the roadway. An enormous log had fallen from a truck. It was too heavy for the villagers to lift, so an elephant had been brought in. Carefully, the massive animal hooked its trunk around the middle of the log, tested the weight, and then raised one end up on to the bed of the truck. Then it backed up and raised the lower end, sliding the whole piece on to the truck. The crowd dispersed, and we drove on through the night.

Several days later, I stopped to visit Ganesh Pahar, a small village situated on the jungle-covered slopes of the Brahmaputra River. It was early morning, and I photographed the fields of tea and rice glistening in the sunlight, the women weavers, and the small children who put their arms around each other and smiled as I took their picture. Many villagers had seen few pictures of themselves, and they laughed to see themselves in the digital monitor. A Hindu shrine to the elephant-headed god, Ganesh, said to be the "remover of all obstacles," was set into the side of the mountain close to the village, the elephant painted bright red and surrounded by offerings.

The best way to see many of the jungle animals is via elephant, since the wild animals ignore the humans who ride them. I set off about 6:00 a.m., with the *majout* (or driver) on a small elephant. We rode through elephant grass so tall it was almost as high as the platform I sat on; it felt like being on an airplane clipping the tops of the clouds. What was hidden in the grass? Though it was not always easy to see beyond the elephant's head, I spotted lots of deer, a pair of tigers curled up together in a nest of grass, and then, finally, one-horned rhinos, which allowed us to approach quite close. They are prehistoric creatures with large plates of skin that look like medieval armor. They eyed us in an unfriendly way but fortunately ignored the elephants. Several had babies, who waddled close to their mothers. In the misty, early-morning light, they seemed like creatures from another time. After I dismounted, I tried to shoot still lifes of the elephants—the details of their eyes, feet, and skin. They seemed like gentle creatures, devoted to their *majout.* A couple of their babies followed through the forest.

Today, India's wildlife refuges are small but vital pockets of preserved vegetation and wildlife that are constantly threatened by encroaching civilization. Of the more than 40,000 tigers that prowled the forests at the time that Kipling's *The Jungle Book* (1894) and *Just So Stories* (1902) were published, only a few thousand remain, despite the government's efforts to create reserves. Their numbers have been reduced because of extensive poaching for their valuable skins. In addition, their teeth, claws, and other body parts are used in Traditional Chinese Medicine (TCM) and exotic recipes, garnering exorbitant prices.

Kipling's jungle is rapidly vanishing and with it the spectacularly beautiful birds, deer, elephants (both wild and tame), monkeys, rhinoceros, and wild boar. Wildlife refuges have helped slow the decline, but their borders are easily breached by poachers, and India's expanding population needs roads and services. In losing the tiger and the jungle, we lose more of the primitive and mysterious wildness that has long been part of our human psyche. Seeing a tiger in the wild is a rare and special gift. I fear that, with their numbers steadily decreasing, it is unlikely that my grandchildren will have the opportunity I had to see a tiger in Kipling's forest.

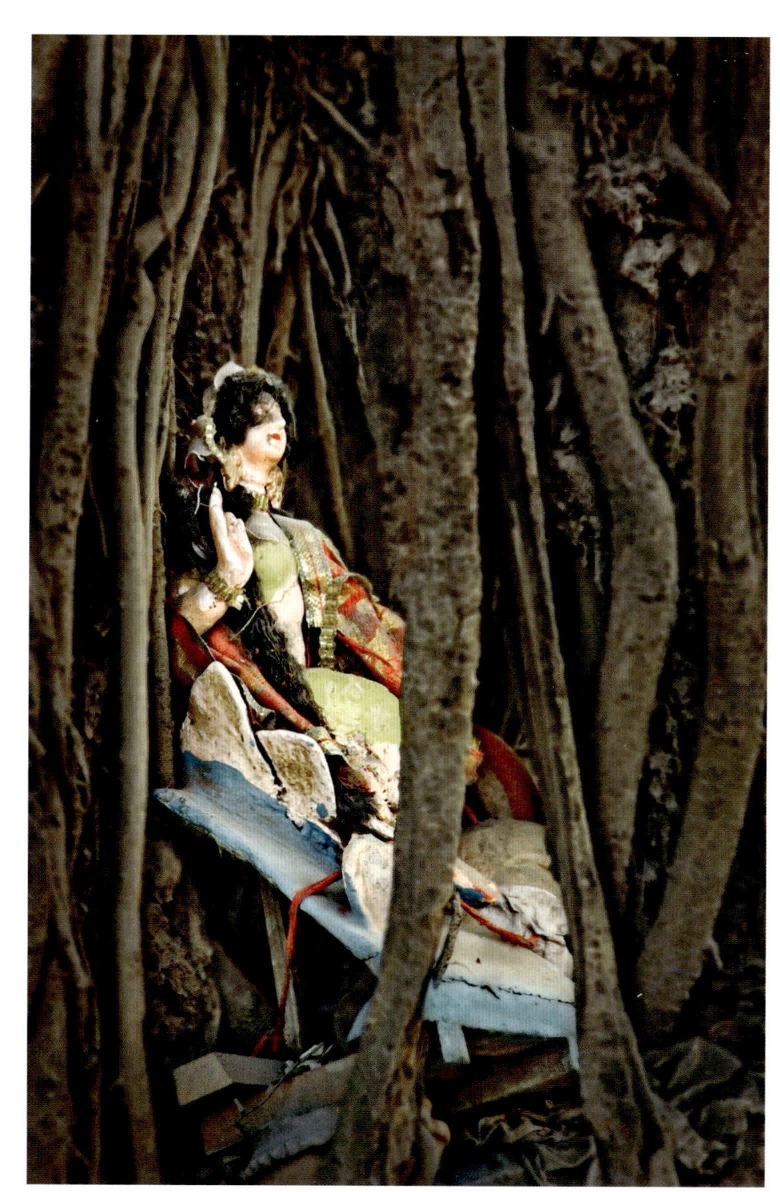

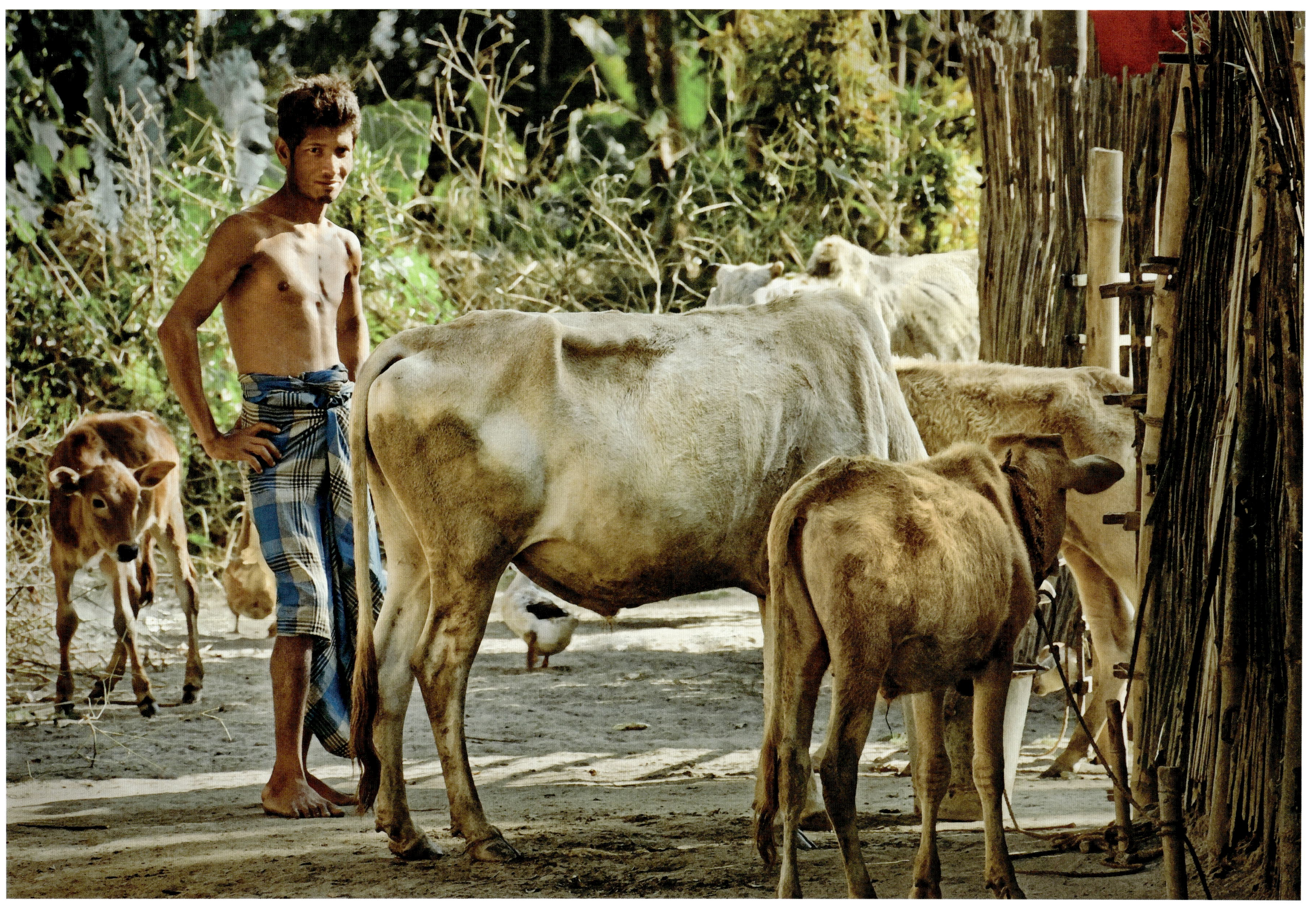

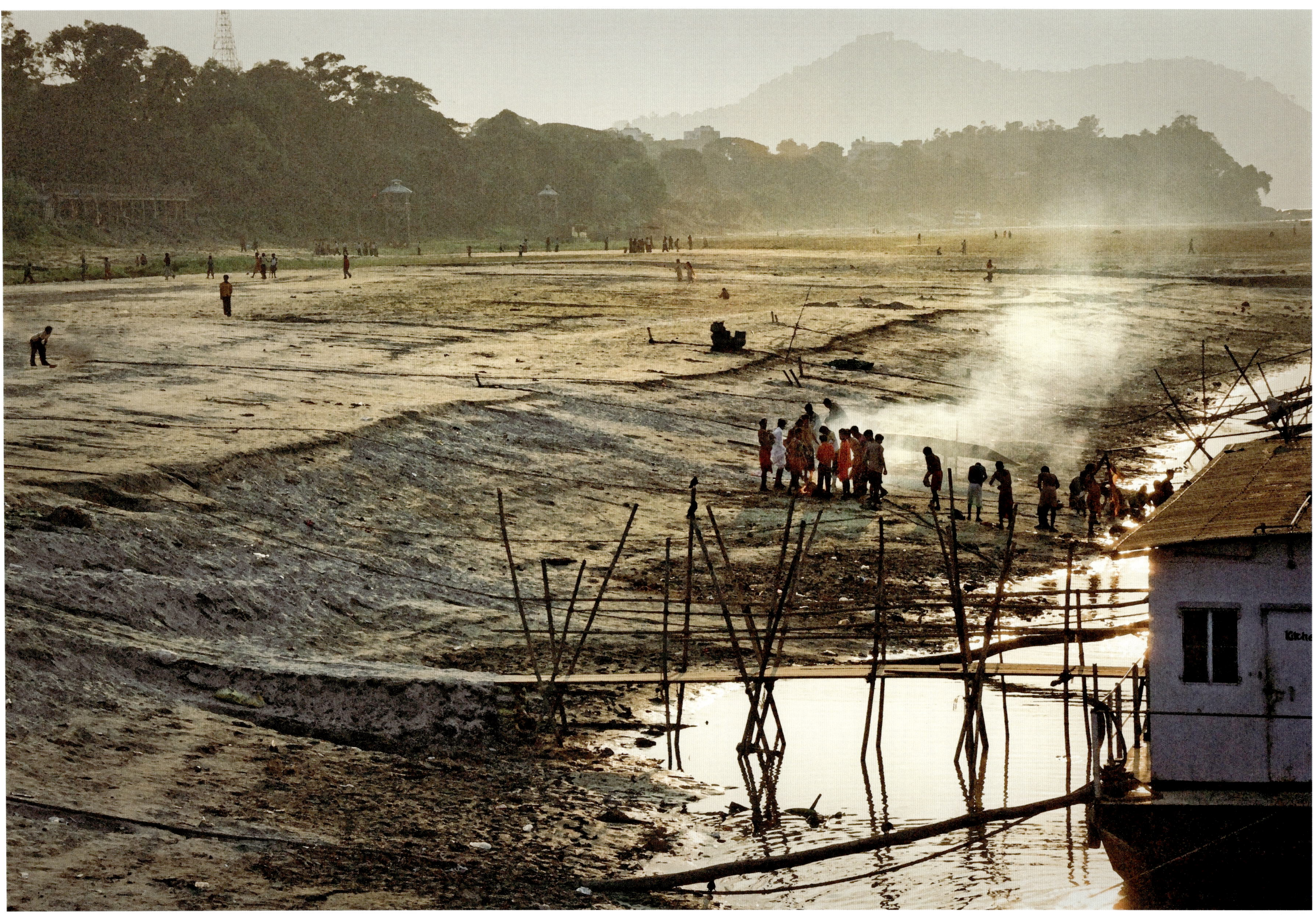

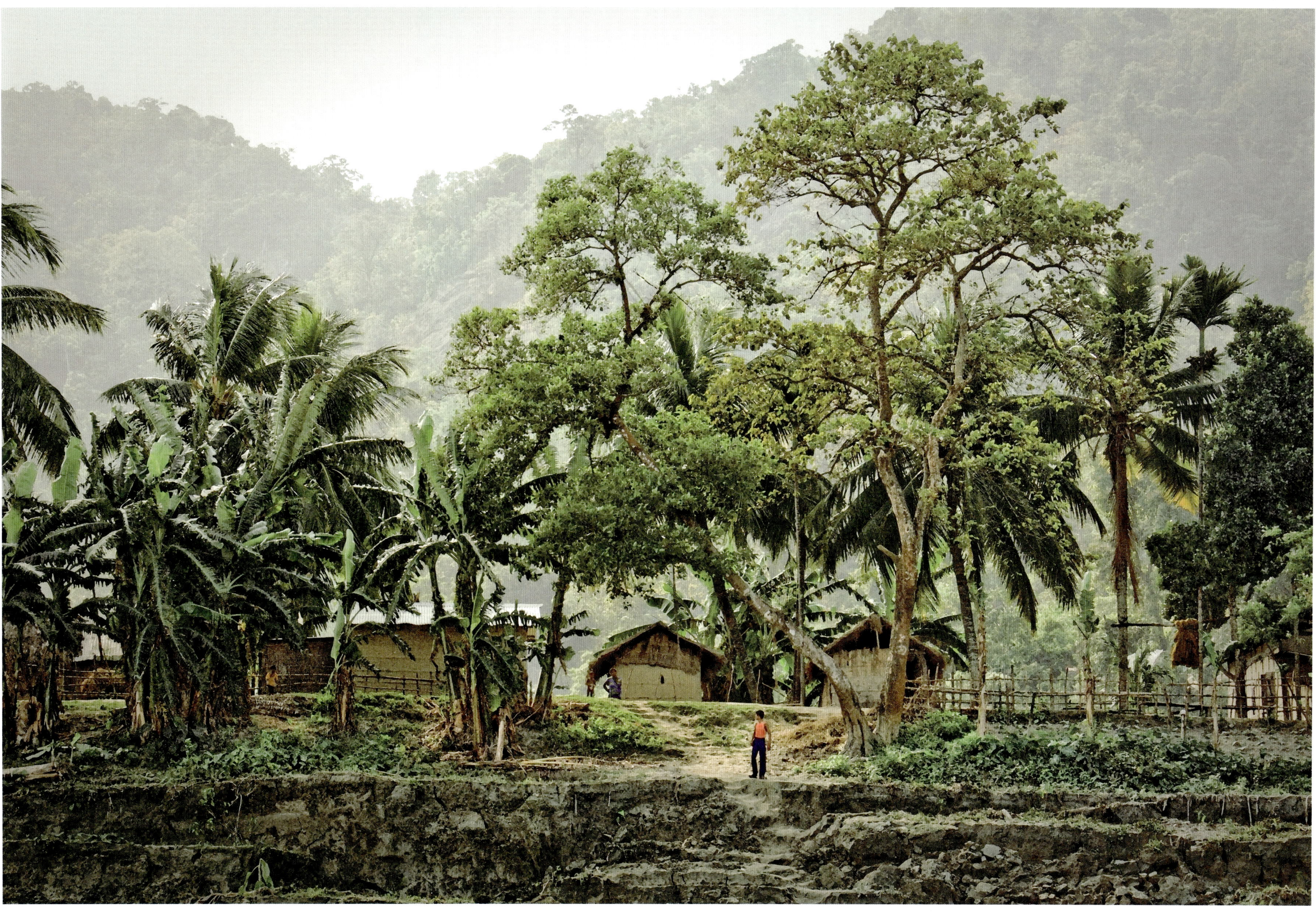

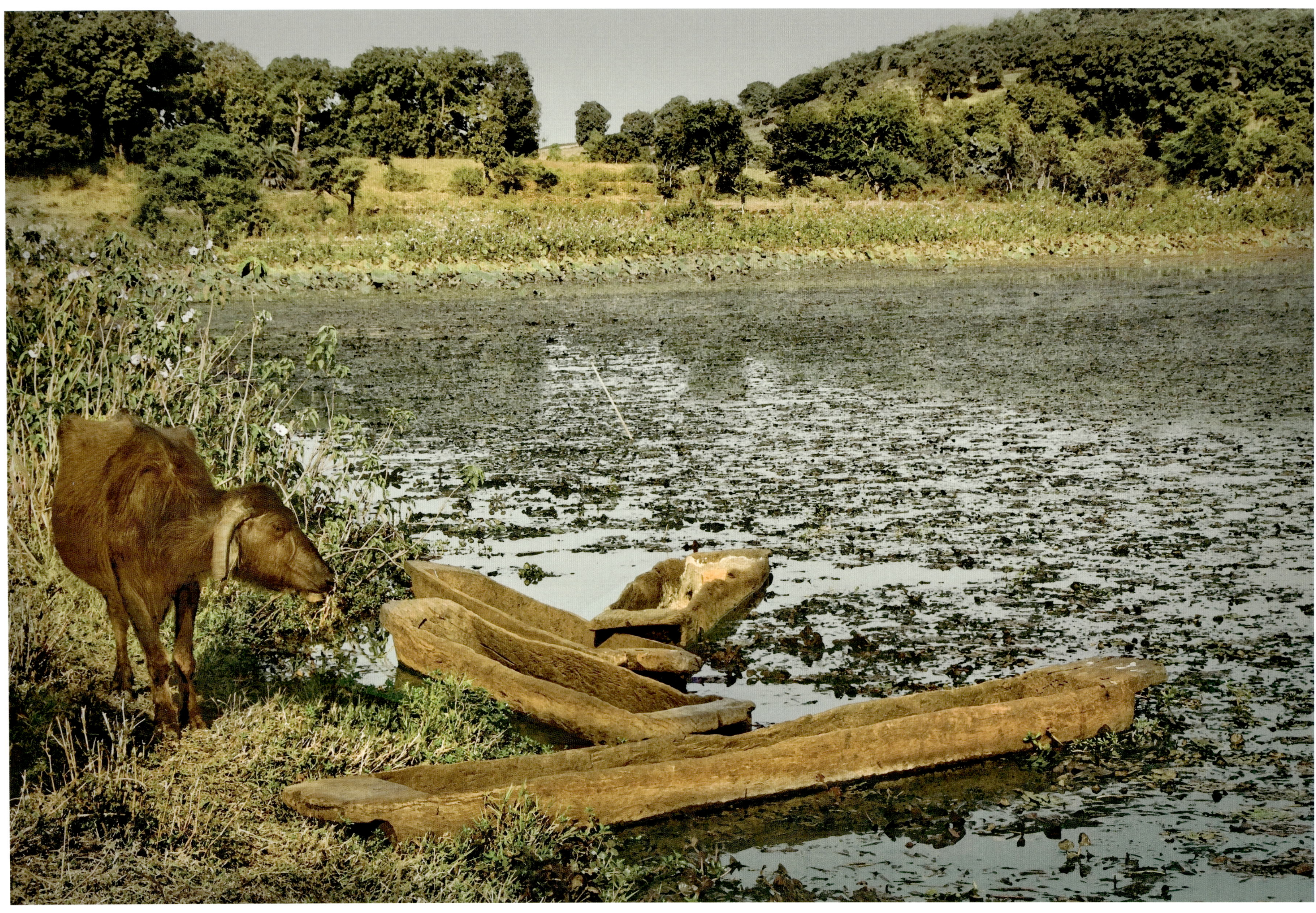

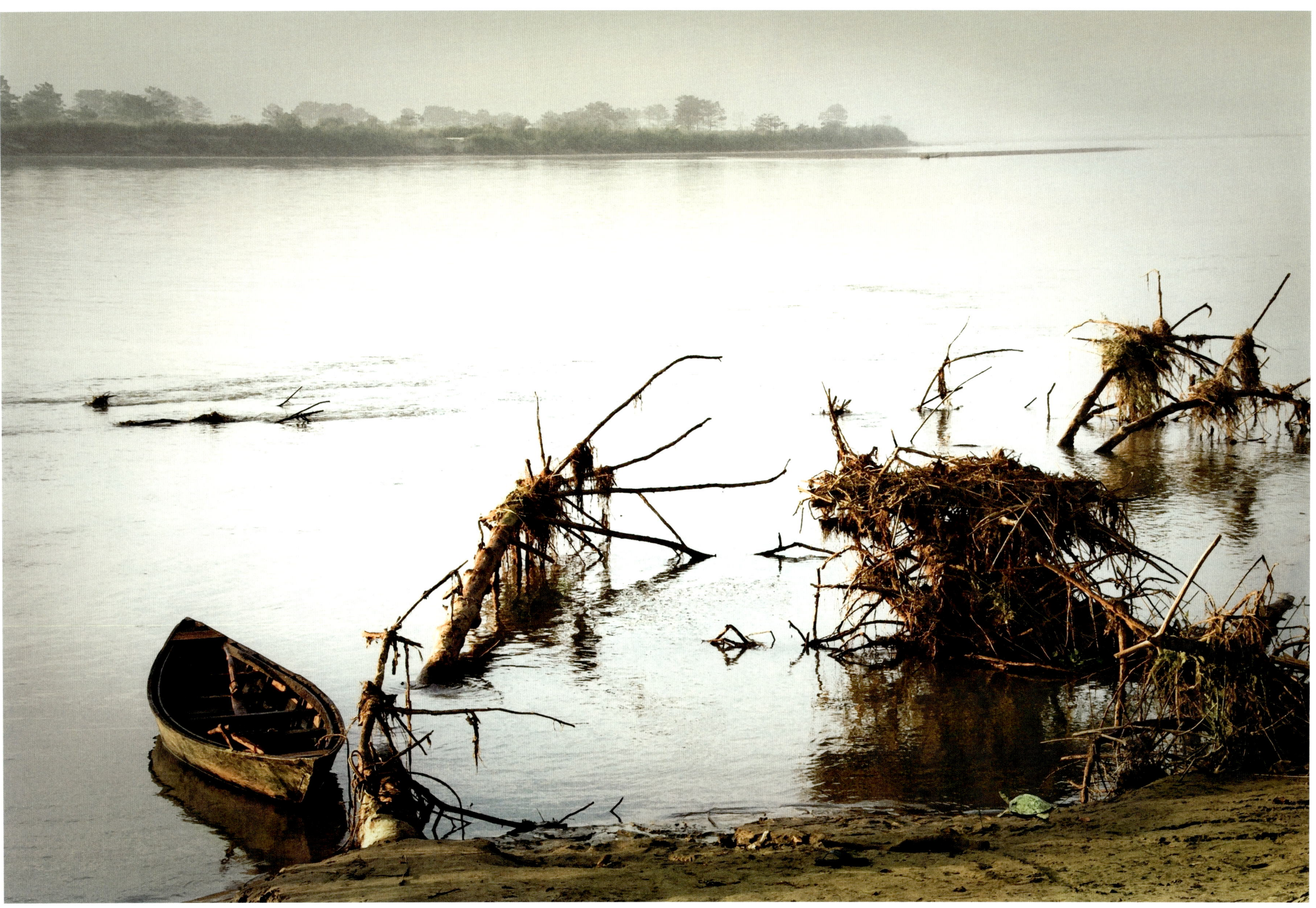

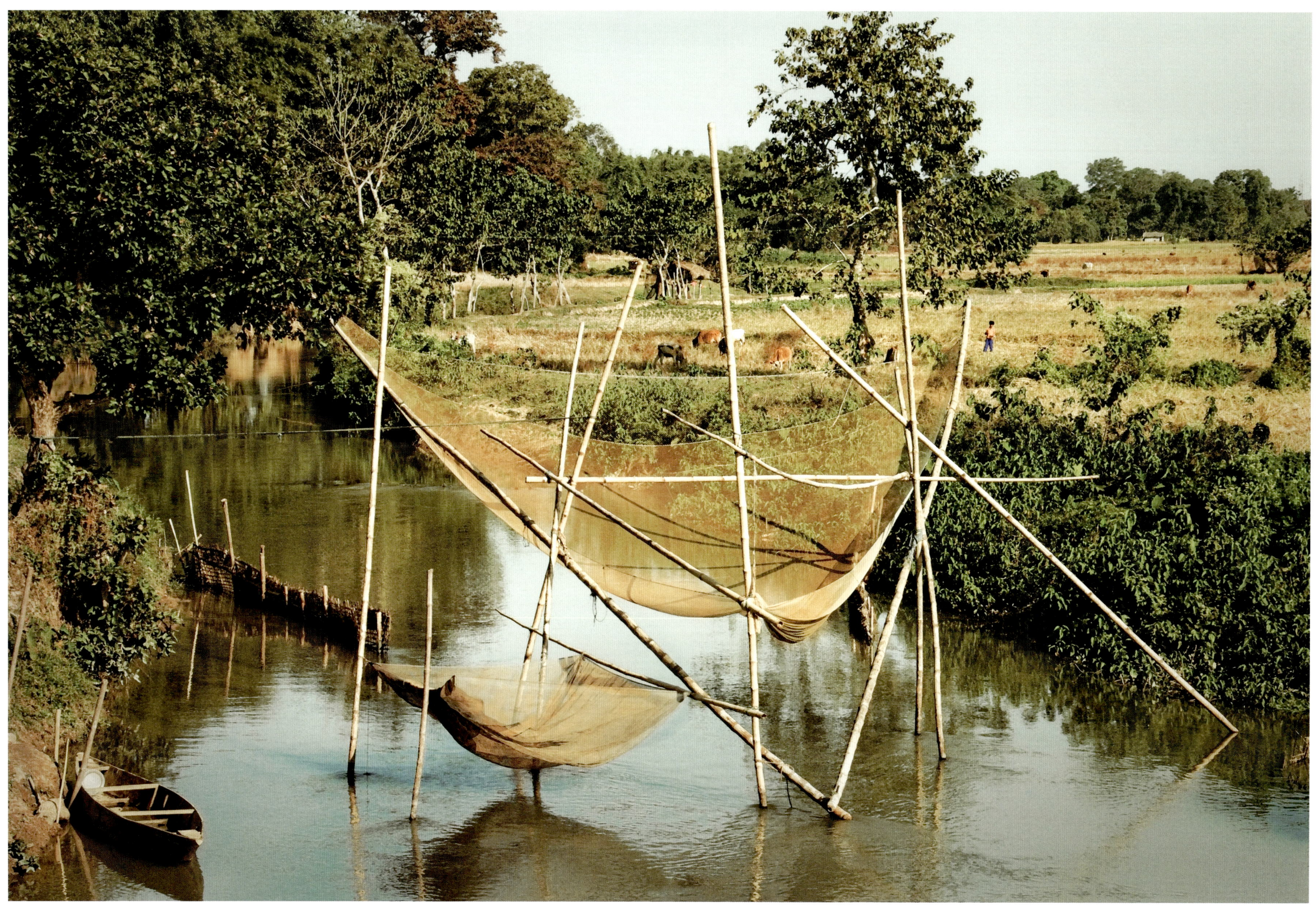

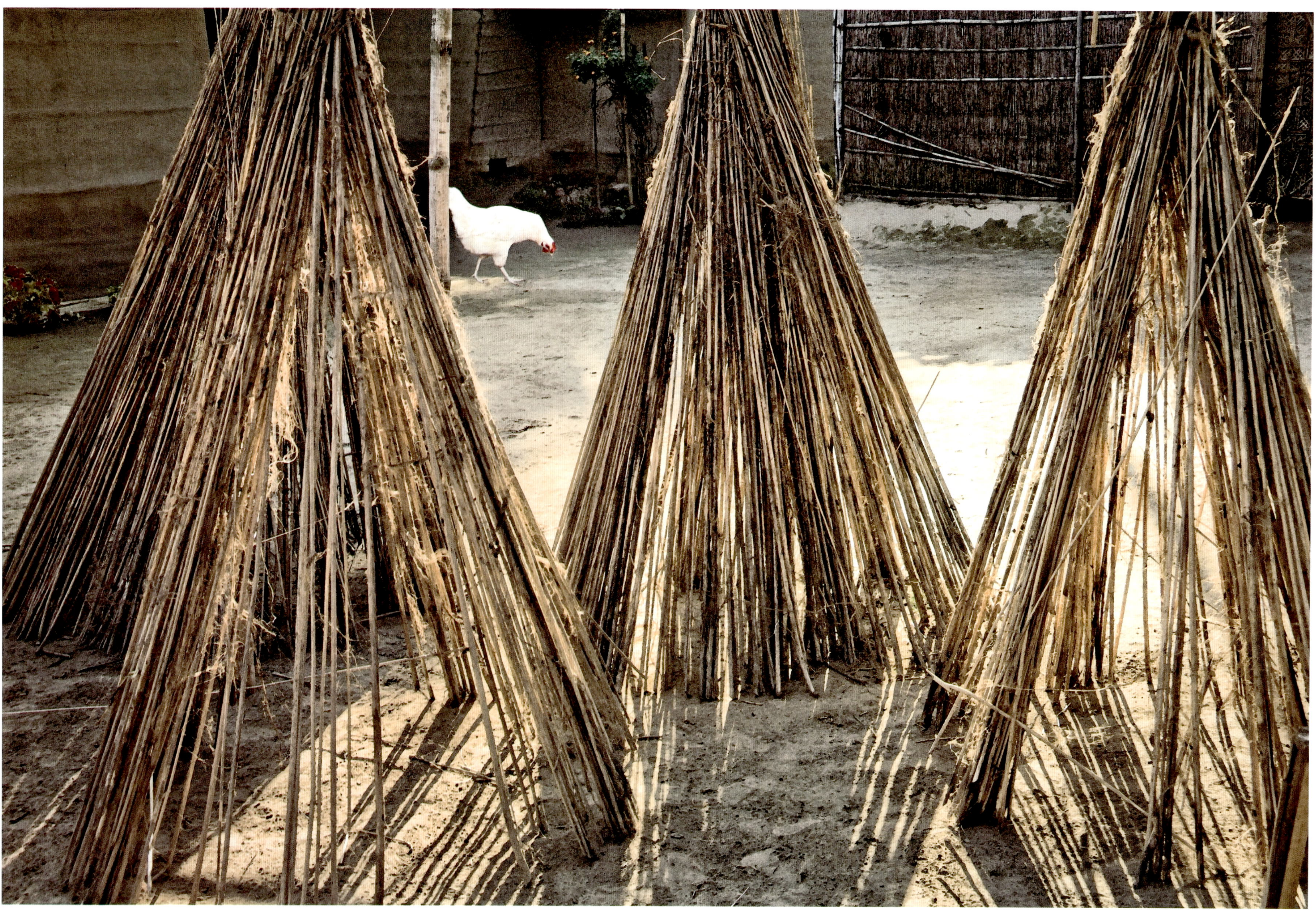

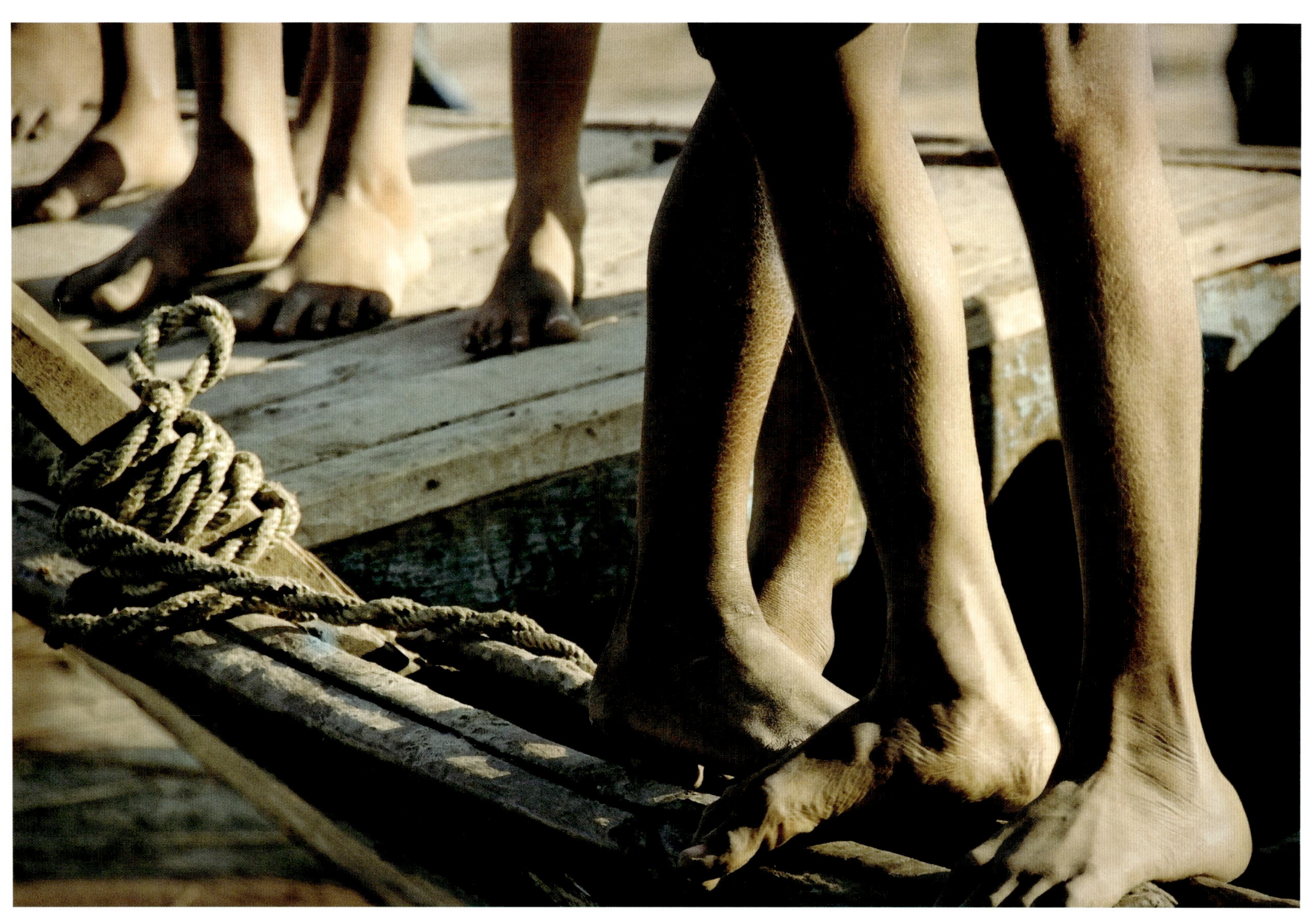

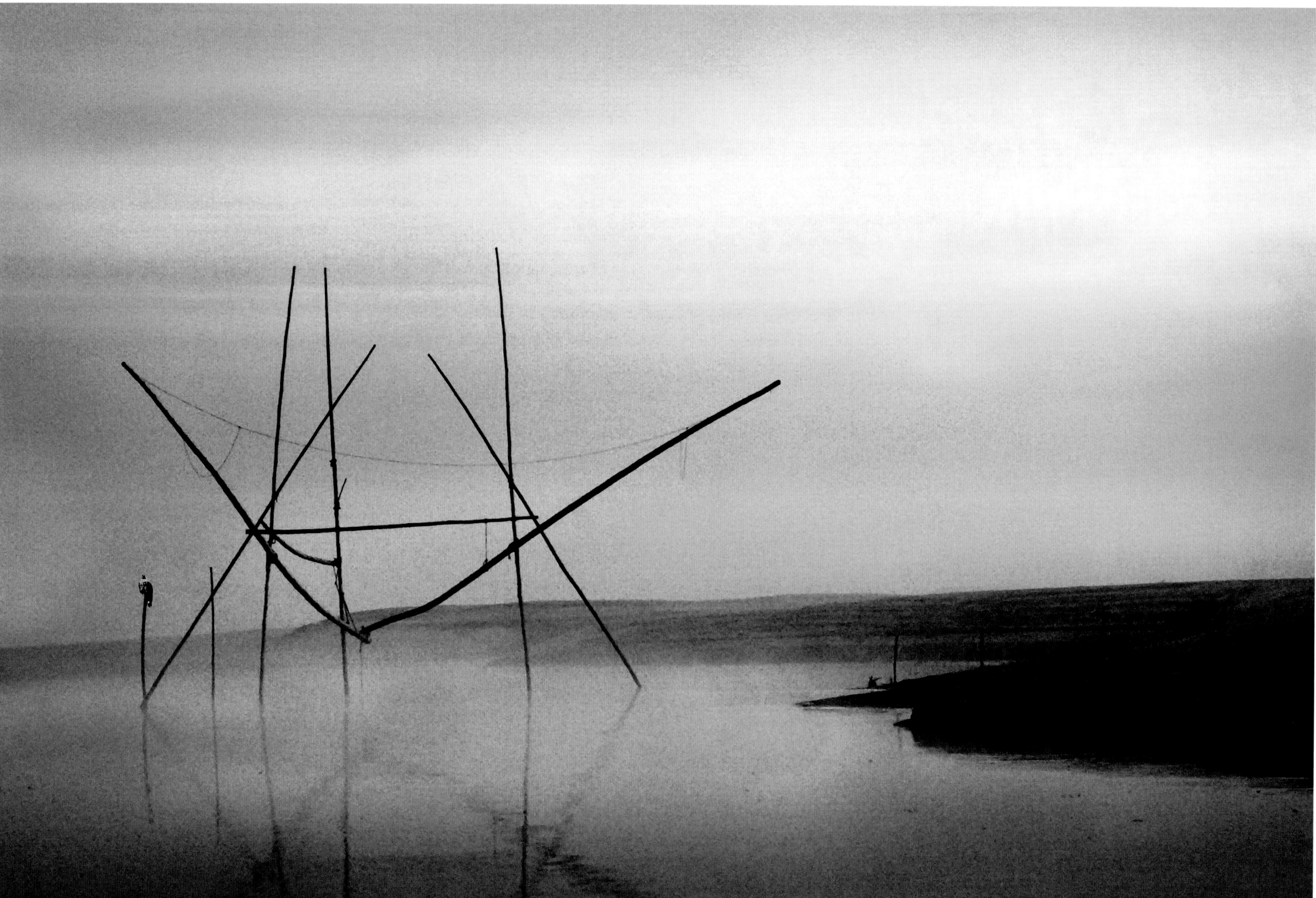

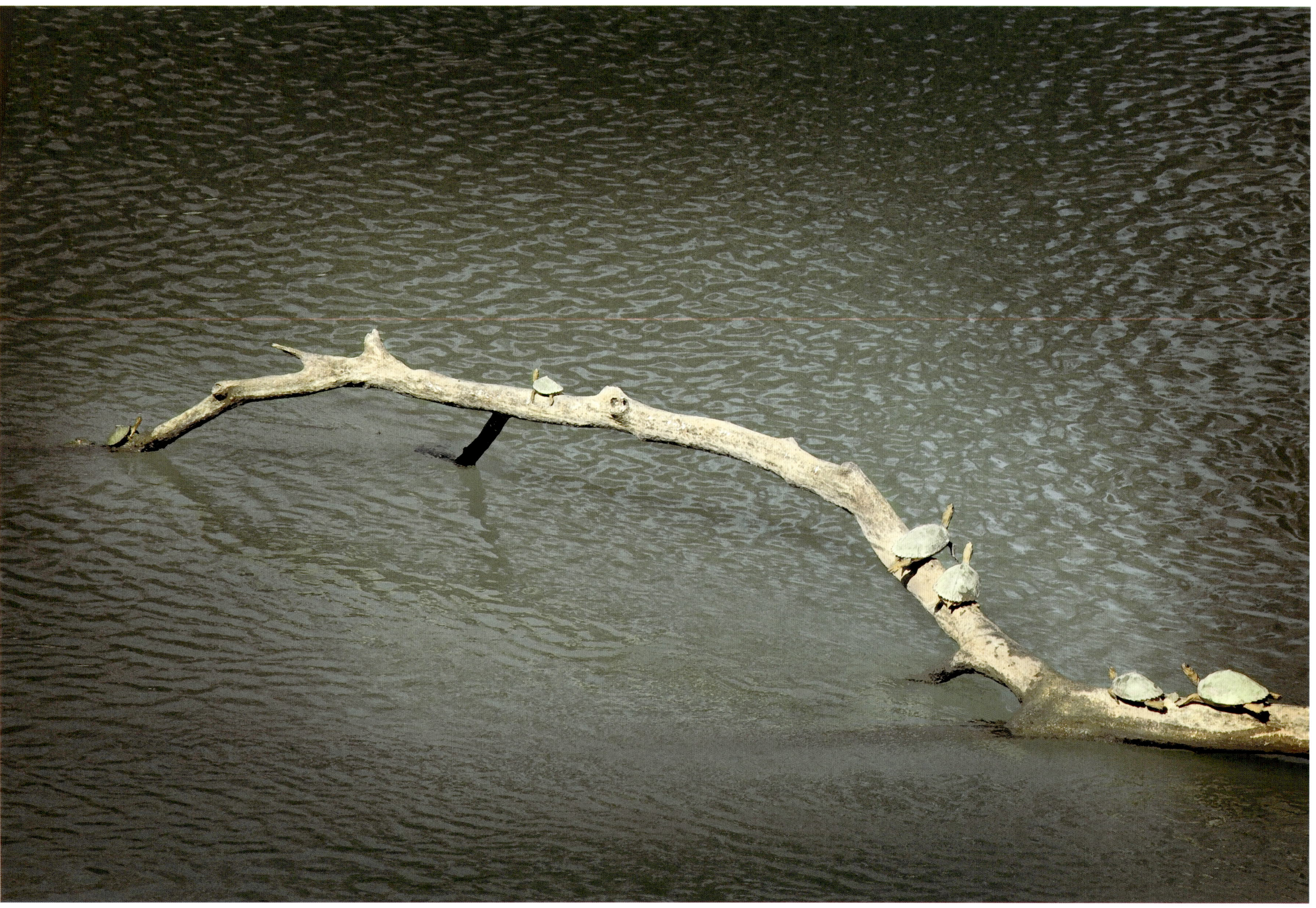

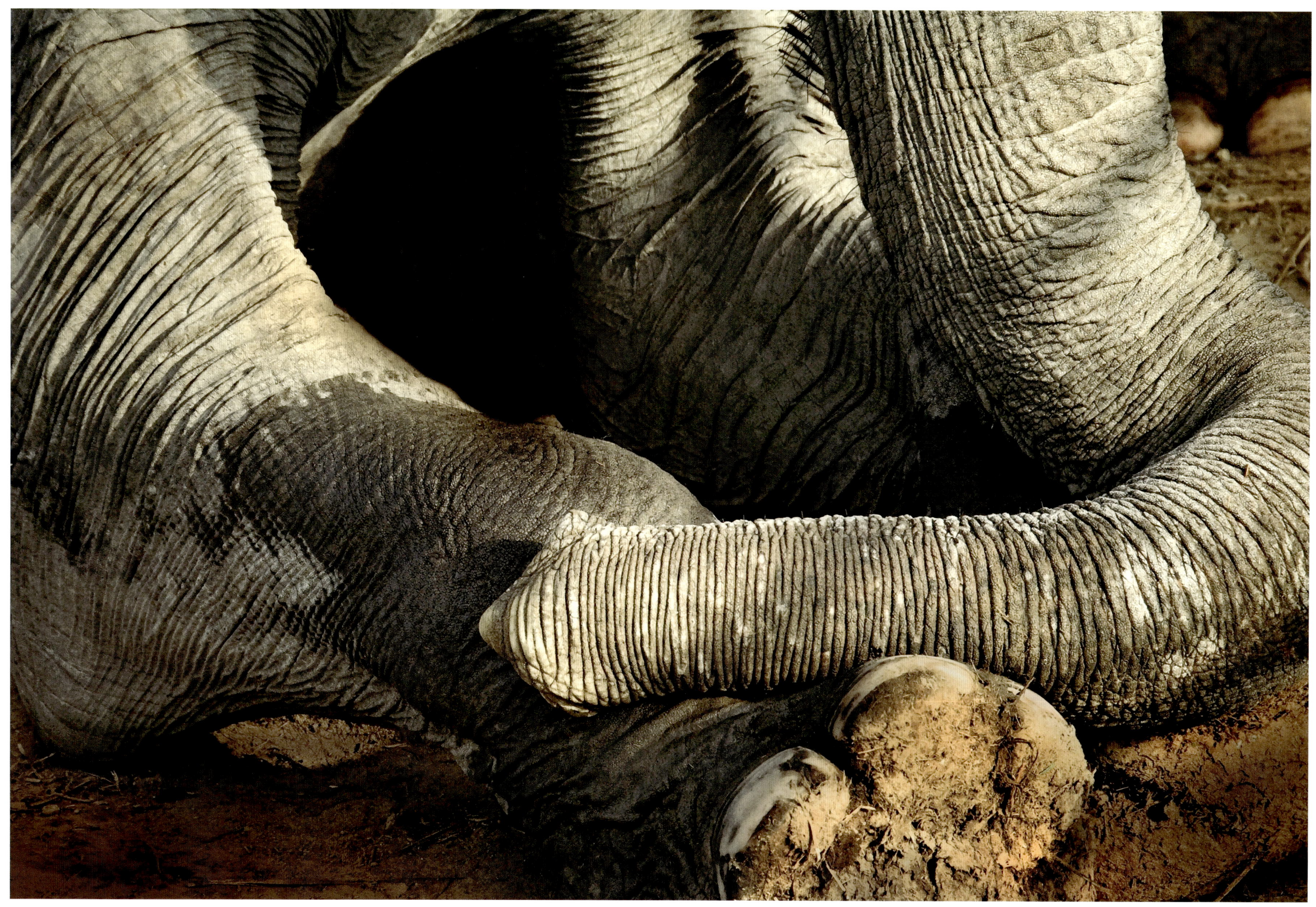

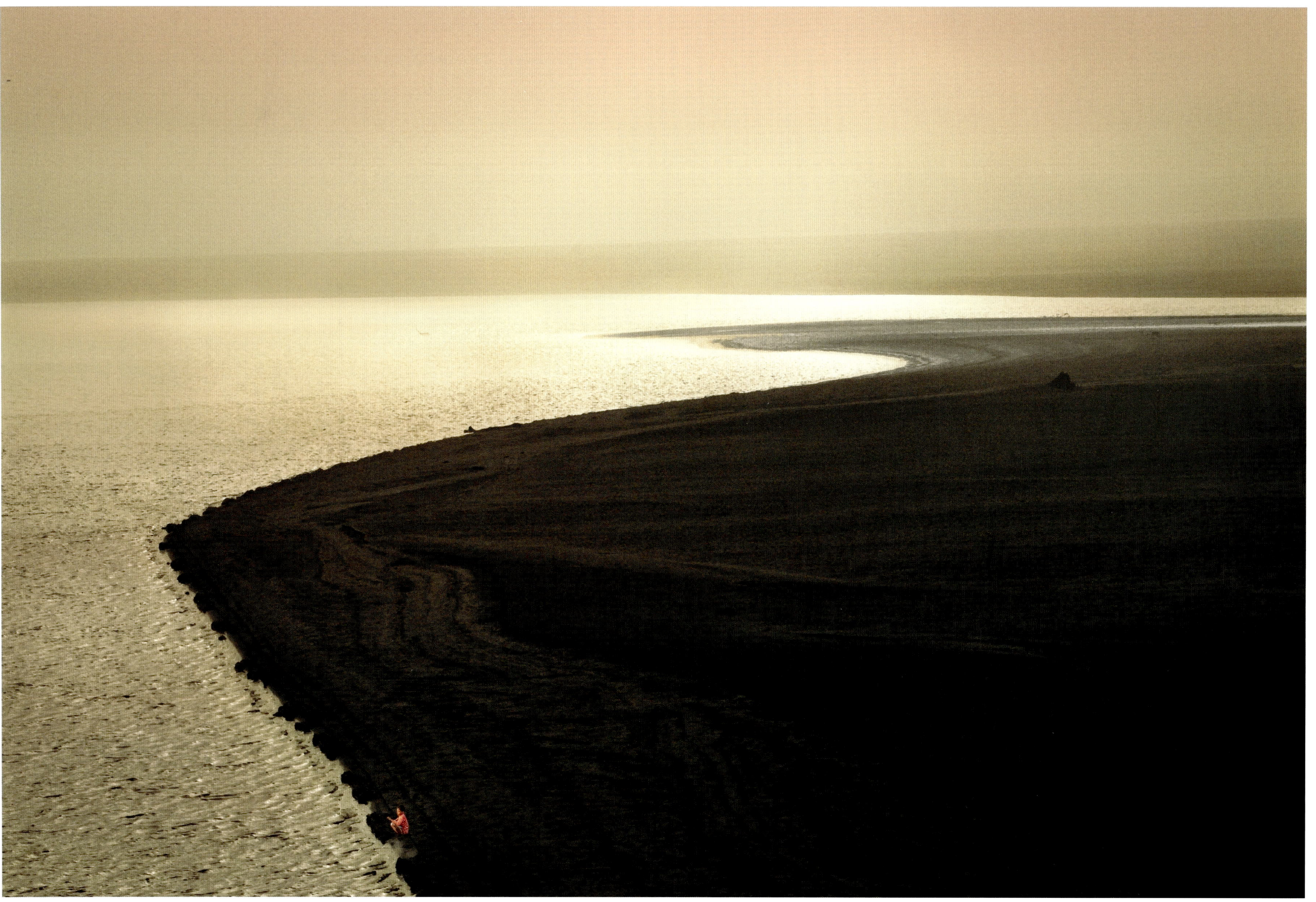

List of Plates

Most of the photographs in this book were taken in 2009
within and close by four national parks: Bandhavgarh and
Kanha (south of Delhi) and Orang and Kaziranga in Assam (far
northeastern India). I also photographed along the Brahmaputra
River from the city of Guwahati to Orang National Park.

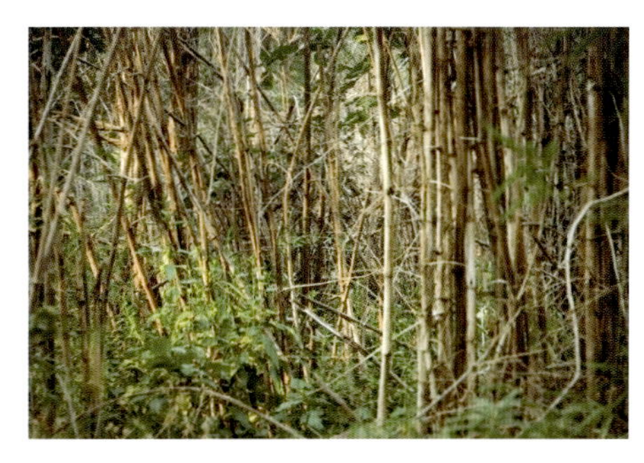

The lush forests and grassland of Kanha National Park, south of Delhi in the Indian state of Madhya Pradesh, provided the inspiration for Rudyard Kipling's *The Jungle Book* and *Just So Stories*.

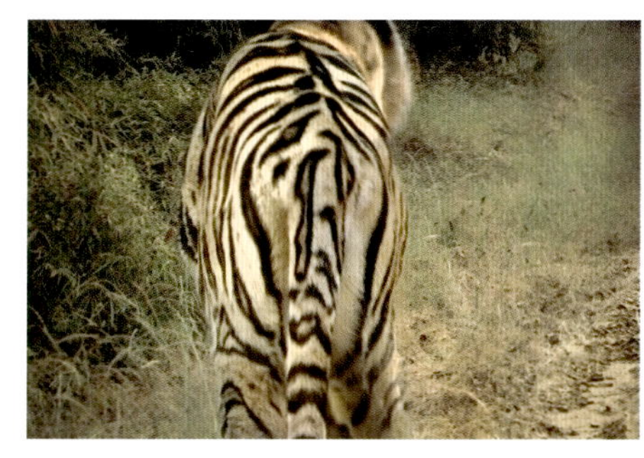

A Bengal tiger (*Panthera tigris*) in Bandhavgarh National Park, southeast of Delhi, which was once the hunting preserve of the Maharaja of Rewa.

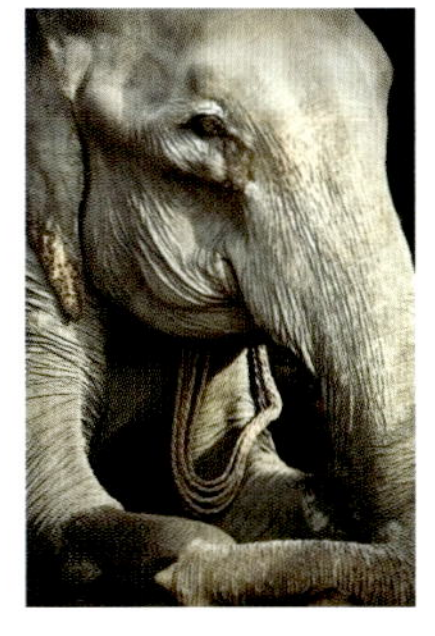

Indian elephants (*Elephas maximus*) are trained to transport tourists in Kaziranga National Park, in the Golaghat and Nagaon districts of Assam in far northeastern India.

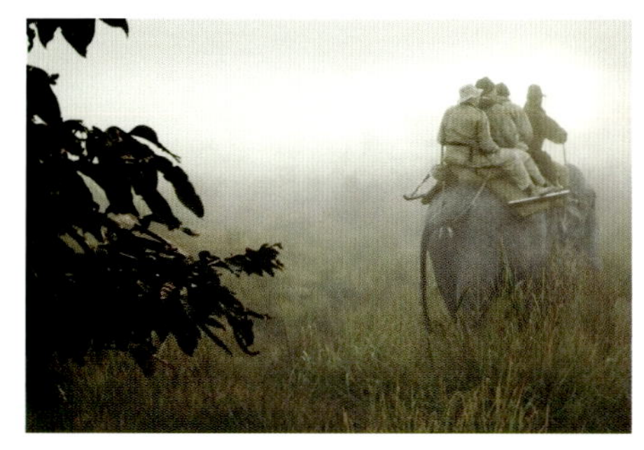

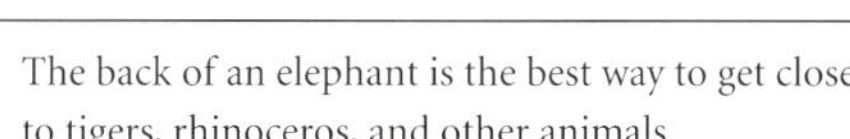

The back of an elephant is the best way to get close to tigers, rhinoceros, and other animals.

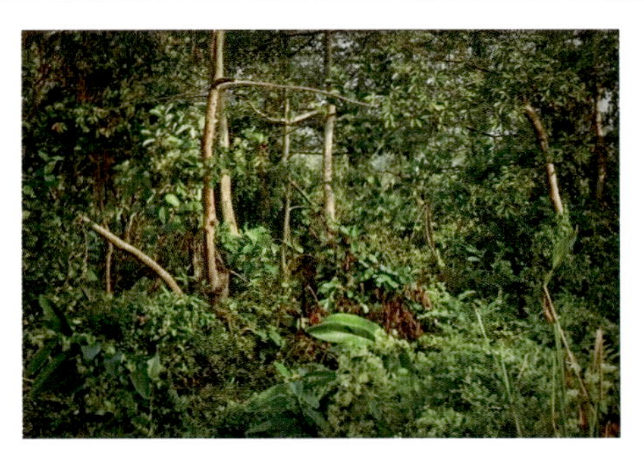

Deep in Kaziranga National Park, one of the world's most important reserves for tigers, one-horned rhinoceros, elephants, deer, water buffalo, and protected birds.

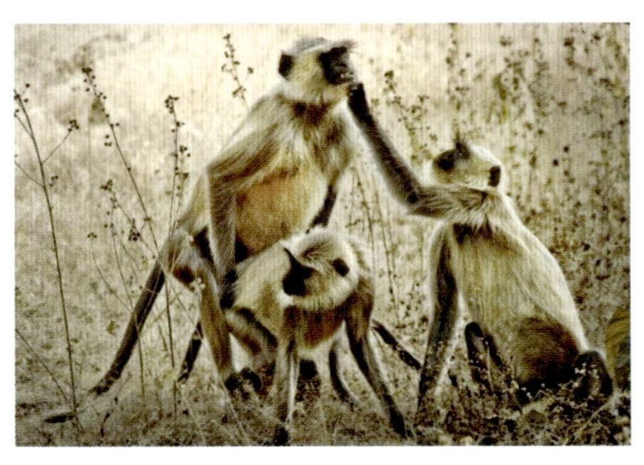

Hanuman langurs (*Semnopithecus entellus*), named after the Hindu monkey-god Hanuman, are the sacred monkeys of India.

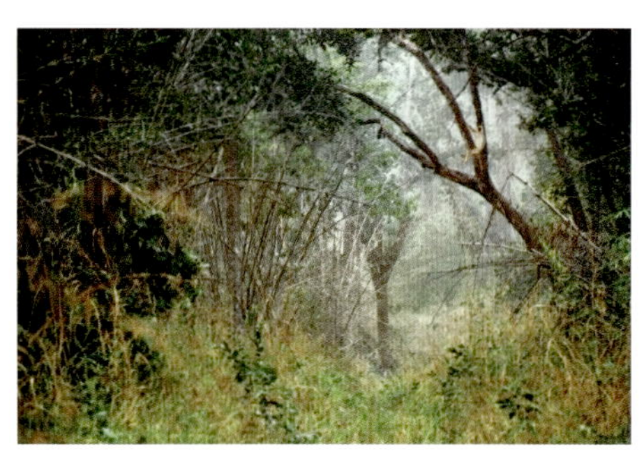

A Hanuman langur in Kanha National Park. Despite laws preventing their capture, langurs are still kept by Hindu priests and sometimes trained for roadside performance.

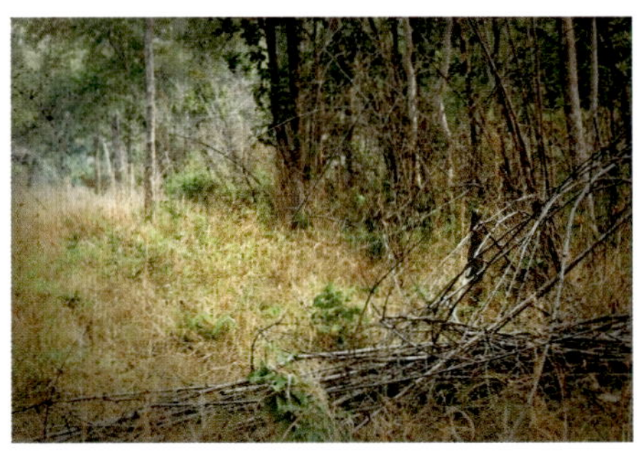

Inside Kanha National Park, established in 1955, home to more than 300 species of birds and large herds of blackbuck, chital, chousingha, sambar, and swamp deer.

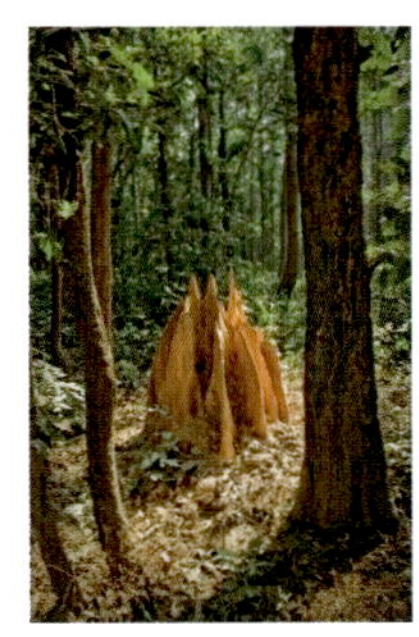

A termite mound in Kanha National Park.

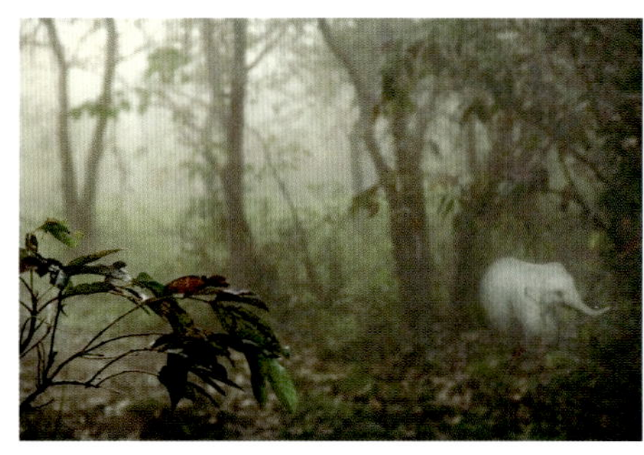

A baby elephant in Kaziranga National Park. Baby elephants are cared for by a tightly knit matriarchal group; it is rare to see a calf without an adult female close by.

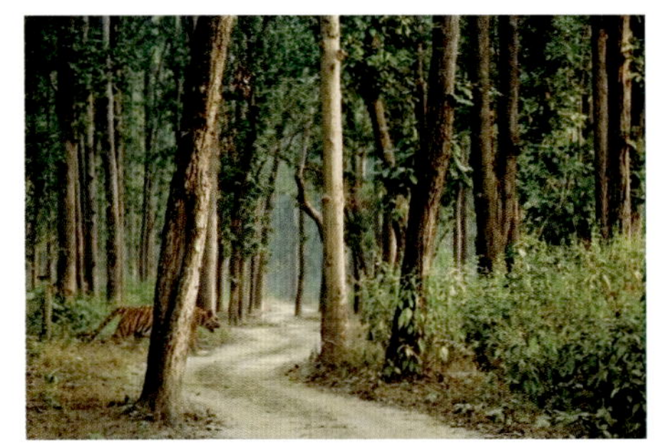

A Bengal tiger in Kaziranga National Park, one of the world's oldest national parks, founded in 1905.

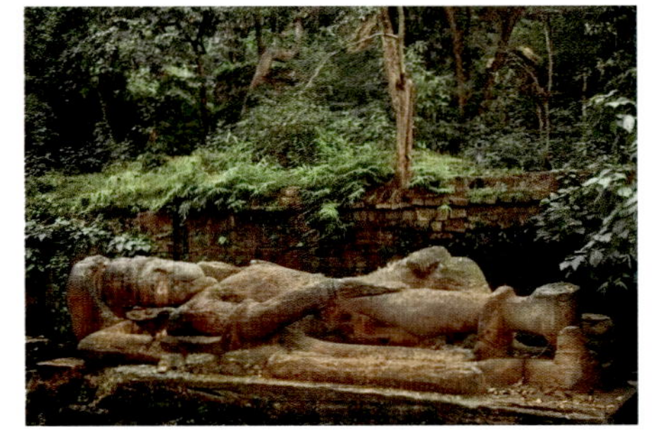

The statue of a reclining Vishnu on a seven-hooded snake was sculpted during the tenth century CE out of a massive outcrop of sandstone in the center of Bandhavgarh National Park.

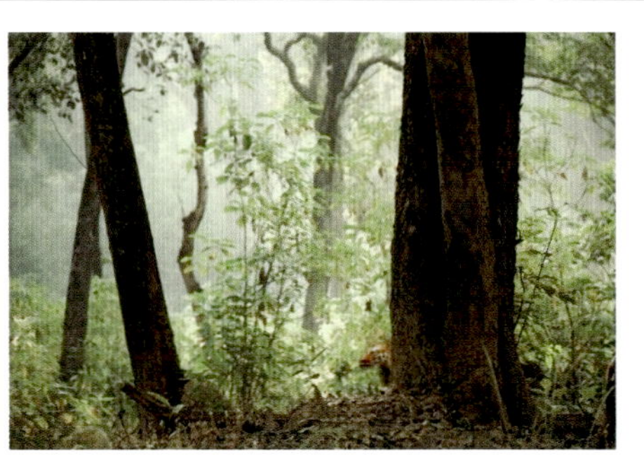

A Bengal tiger in Kanha National Park.

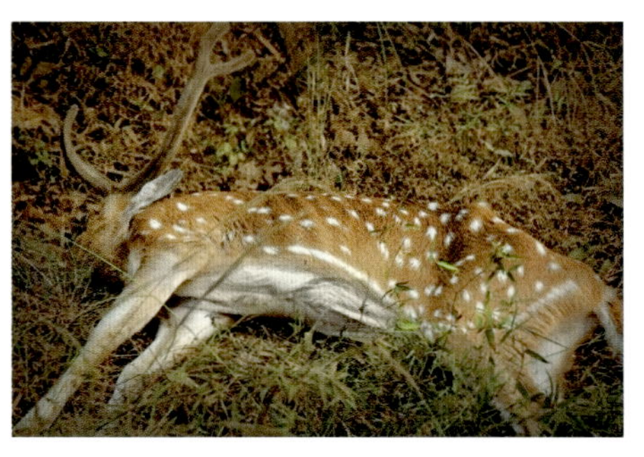

A recent tiger kill of a spotted chital deer (*Axis axis*) in Bandhavgarh National Park.

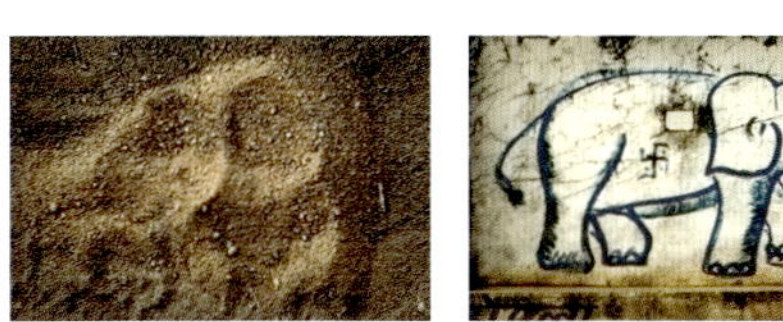

left: A tiger footprint in Bandhavgarh National Park.
right: A contemporary mural in the town of Khajuraho.

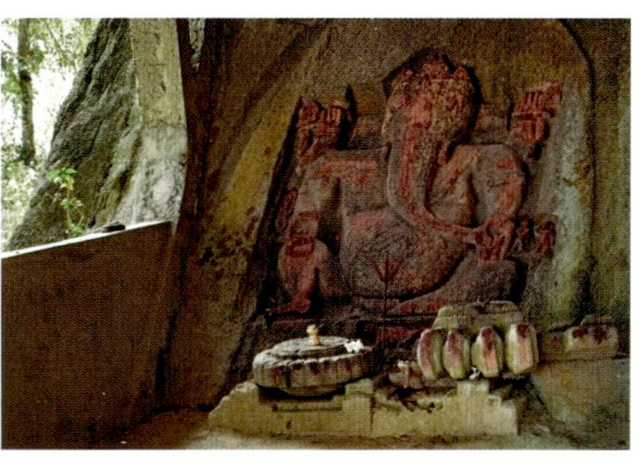

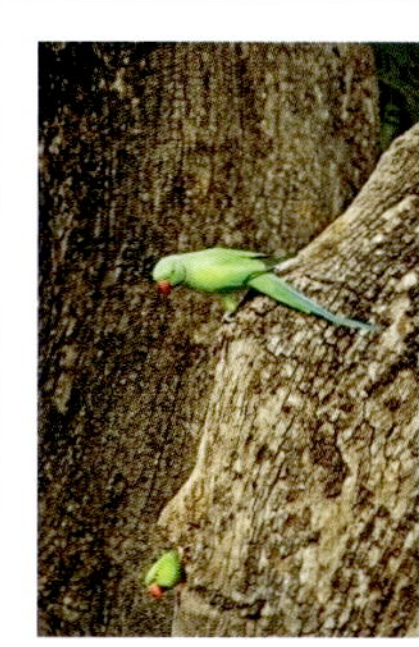

A shrine outside Ganesh Pahar, a small village along the Brahmaputra River.

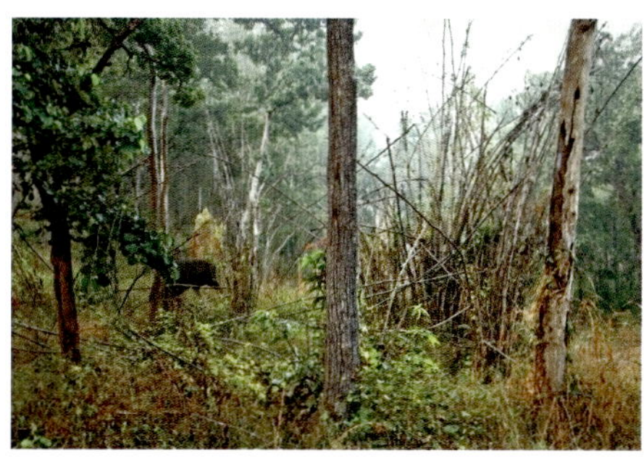

A rose-ringed Parakeet (*Psittacula krameri*) at Khajuraho, a UNESCO World Heritage Site considered one of the seven wonders of India.

A wild boar (*Sus scrofa*) in Kanha National Park.

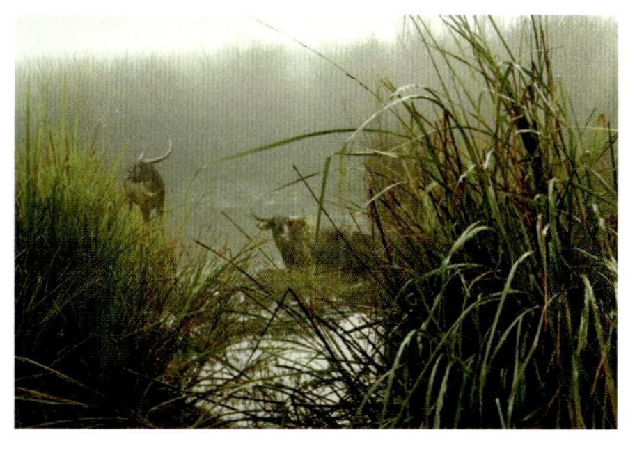

An Asiatic wild water buffalo (*Bubalus arnee*) in Kaziranga National Park. The species has been classified as endangered.

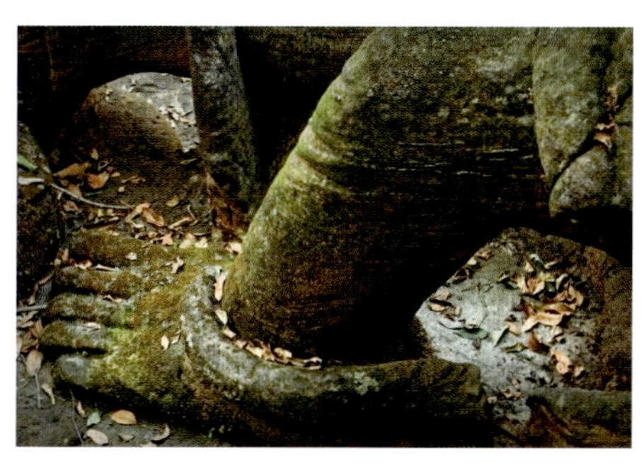

Foot of a sculpture of Vishnu in Bandhavgarh National Park.

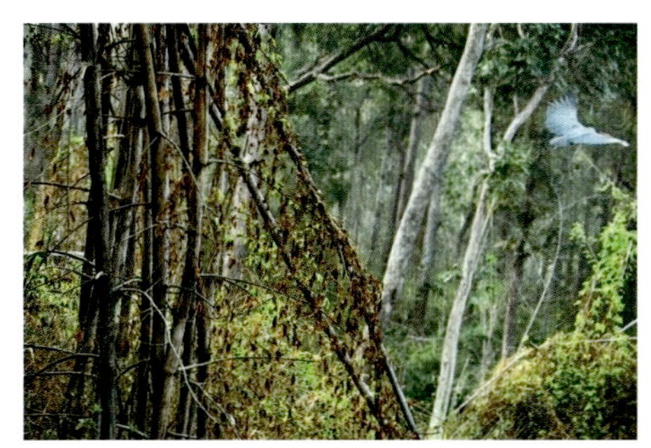

A lesser adjutant stork (*Leptoptilos javanicus*) in Kanha National Park. This species has been classified as vulnerable.

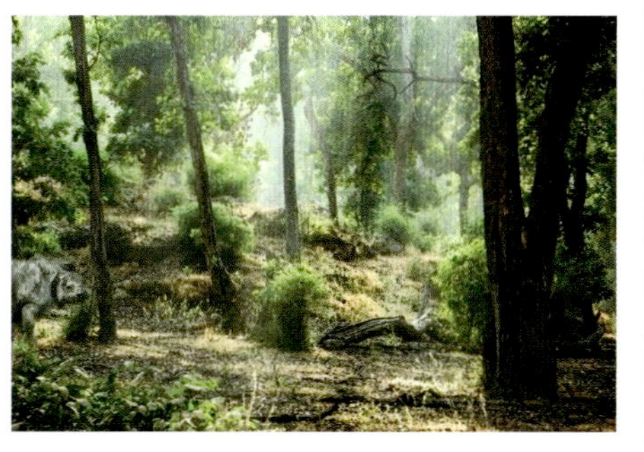

Indian rhinoceros (*Rhinoceros unicornis*) in Kaziranga National Park. This species has been classified as endangered.

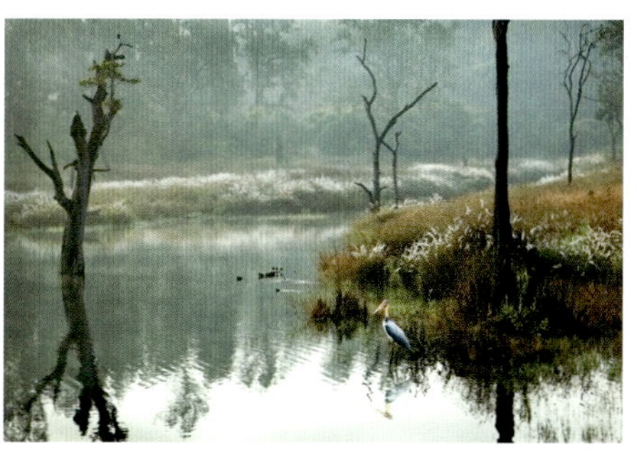

A lesser adjutant stork in Kanha National Park.

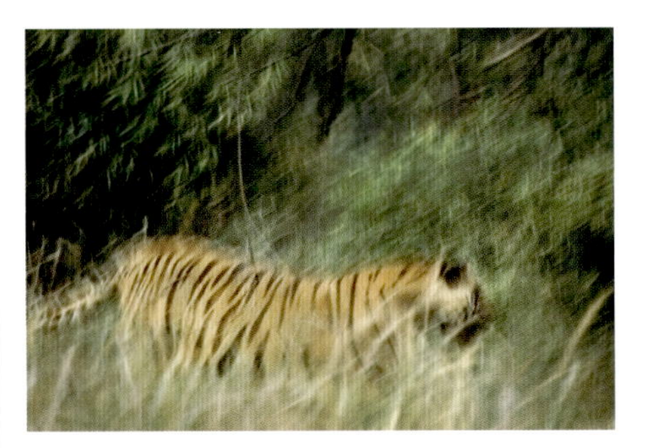

A Bengal tiger in Bandhavgarh National Park.

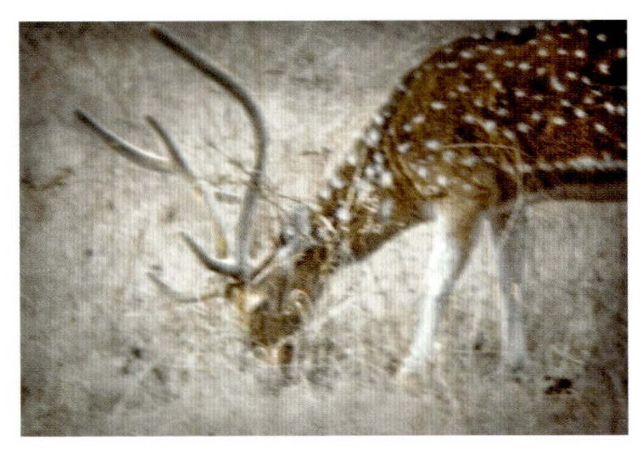

A spotted chital deer (*Axis axis*) in Kaziranga National Park. Chitals are the most common deer in Indian forests.

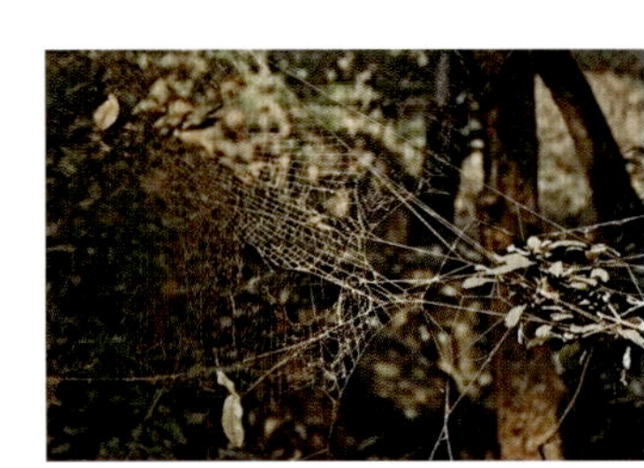

A giant spider web in Bandhavgarh National Park.

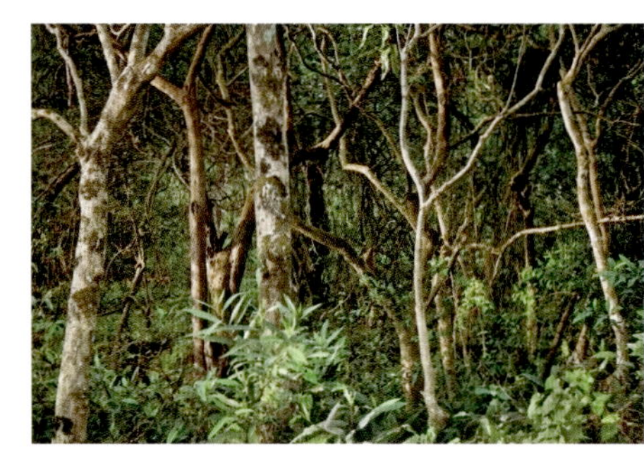

Kaziranga National Park.

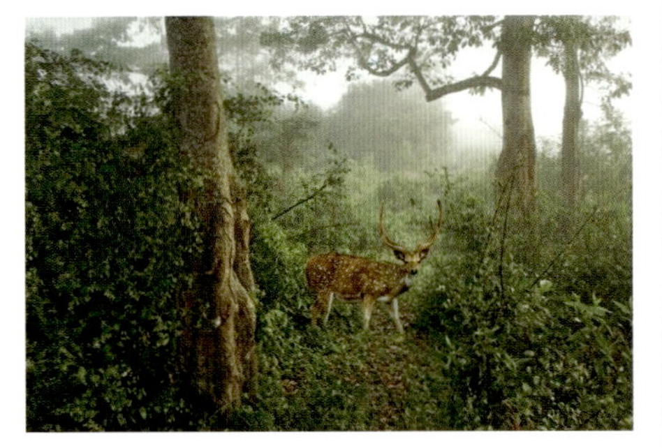

A spotted chital deer in Kaziranga National Park.

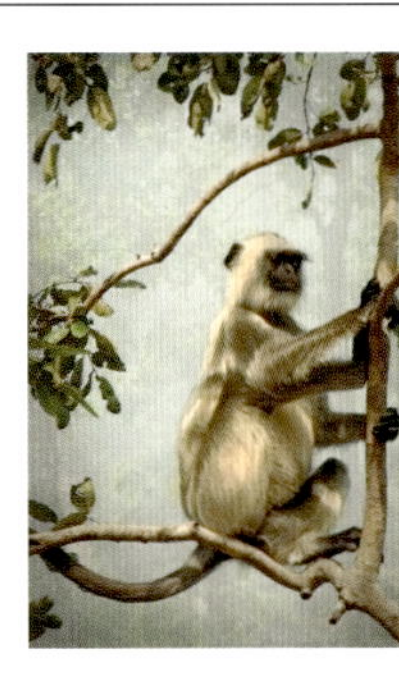

A hanuman langur in Bandhavgarh National Park.

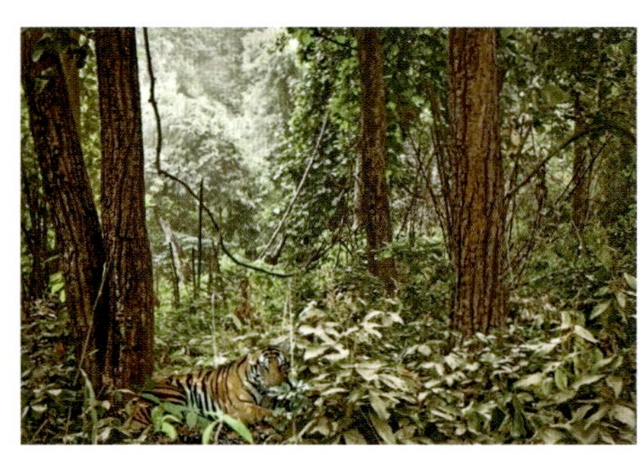

A Bengal tiger in Kanha National Park.

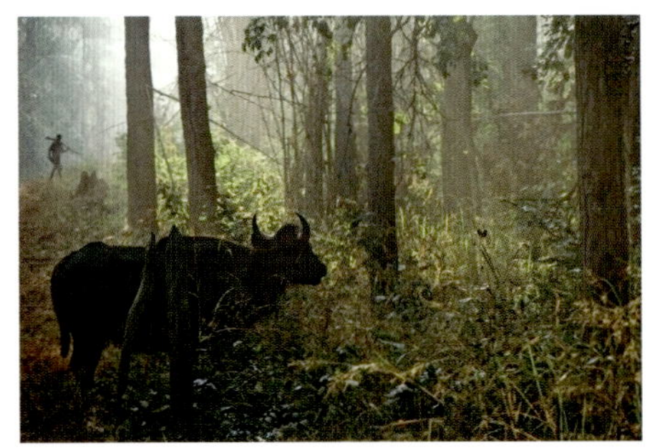

The gaur (*Bos gaurus*) in Kaziranga National Park is a bison-like animal, the largest and most powerful of the bovids, and classified as vulnerable.

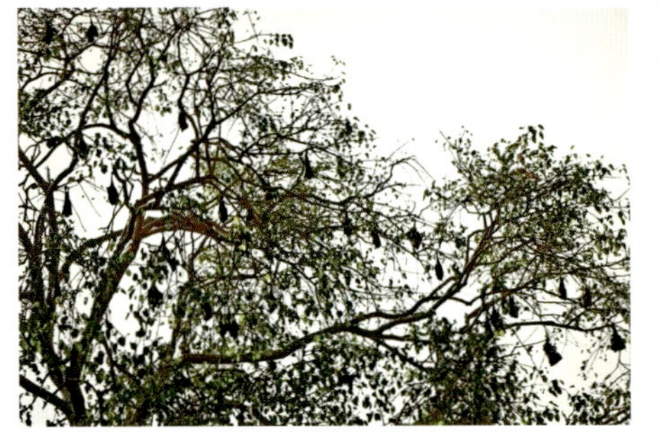

Giant fruit bats (*Pteropus giganteus*), also called flying foxes, are social animals that roost in large, noisy colonies, as shown here in Bandhavgarh National Park.

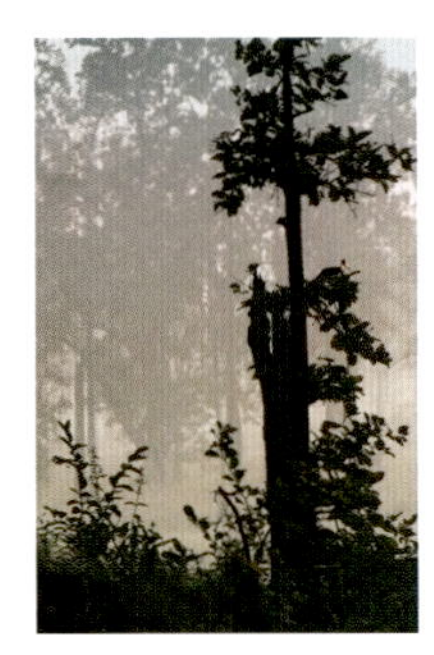

Kanha National Park.

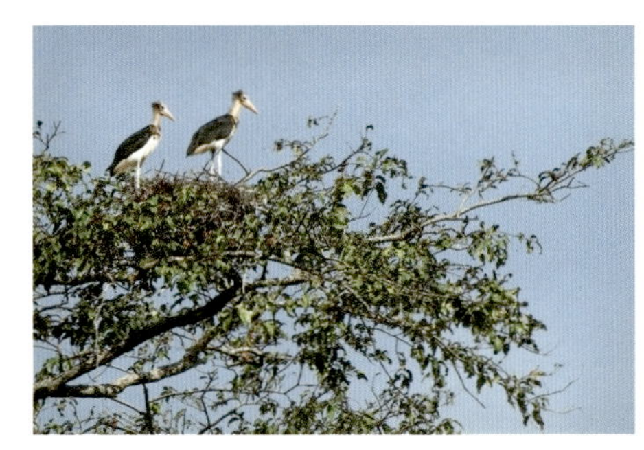

Lesser adjutant stork nests in Kaziranga National Park.

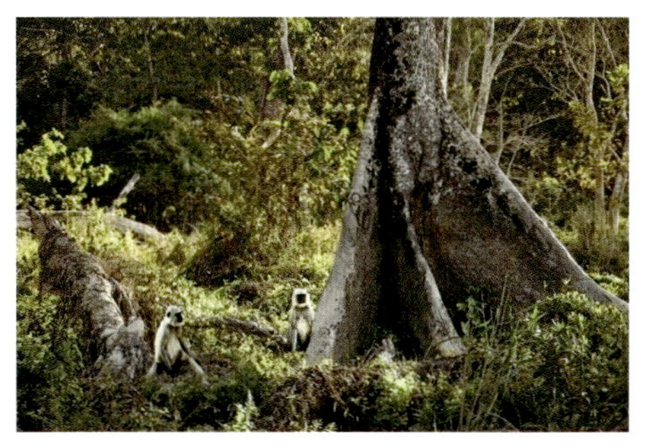

Hanuman langurs in Kaziranga National Park.

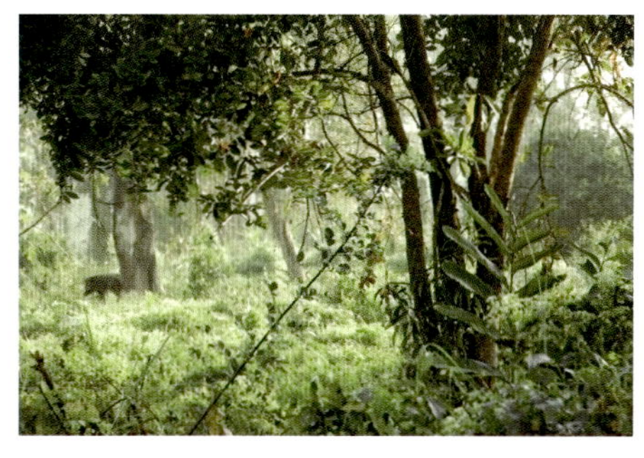

A wild boar in Kaziranga National Park.

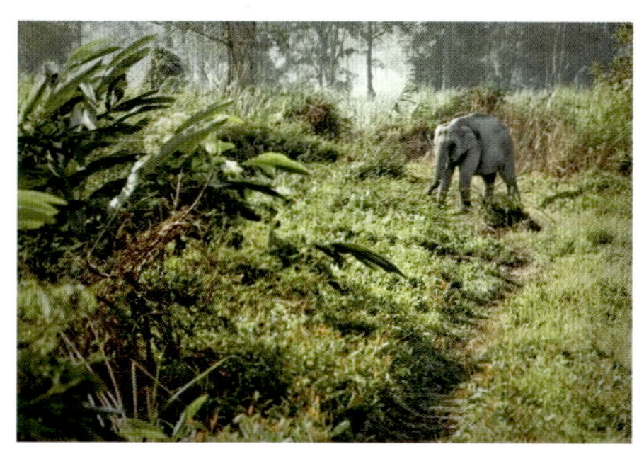

A wild Indian elephant in Kaziranga National Park. Smaller than the African species and classified as endangered, they are more easily tamed and can still be seen lifting heavy logs in rural India.

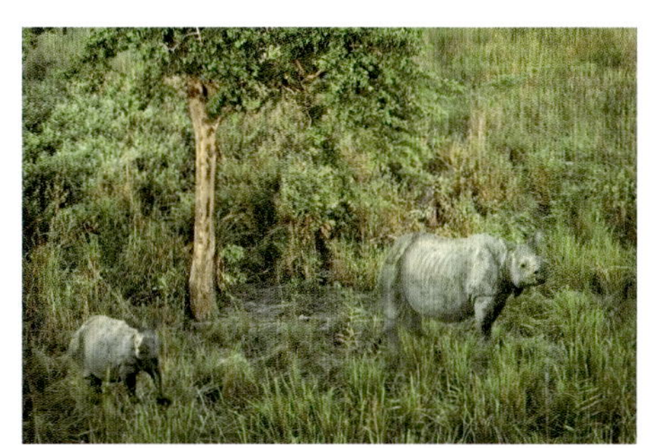

An Indian rhinoceros in Kaziranga National Park. Rhinos are unpredictable, dangerous, and able to charge at more than twenty-five miles per hour.

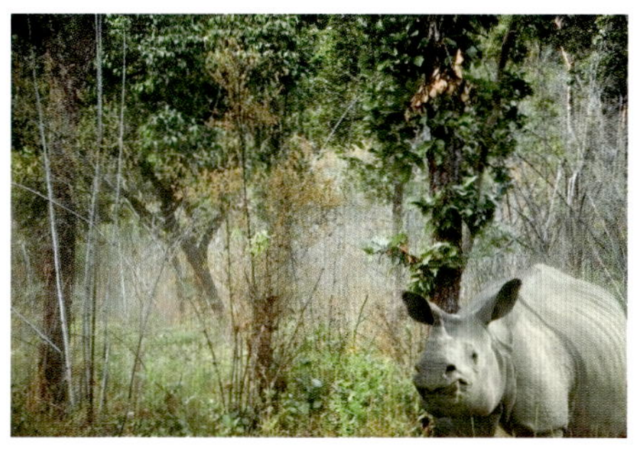

The poaching of rhinoceros is big business despite the efforts of hundreds of armed guards in Kaziranga National Forest. A single-horn rhino can bring more than $30,000 (U.S.) on the black market in Asia for its purported medicinal powers.

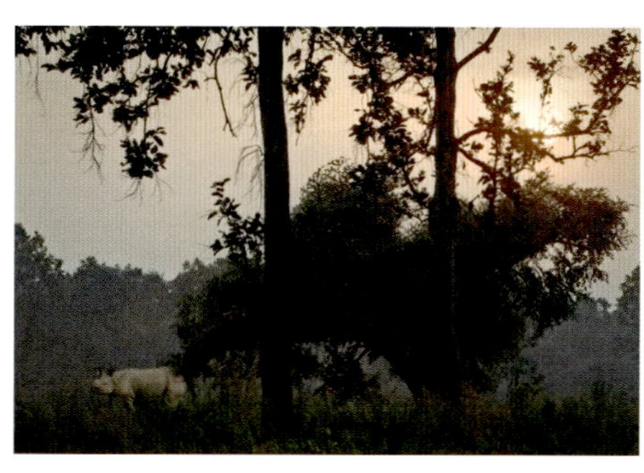

Kaziranga is Asia's premier rhino reserve, but modern development crowds the animals. The park is now encircled by highways, rice fields, small villages, and herds of cattle.

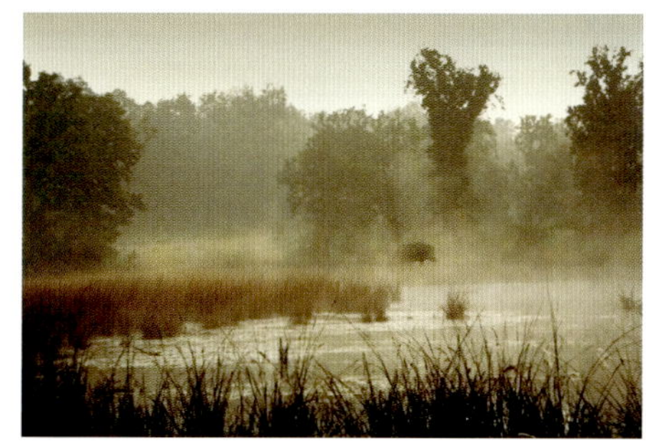

A wild boar in Kanha National Park

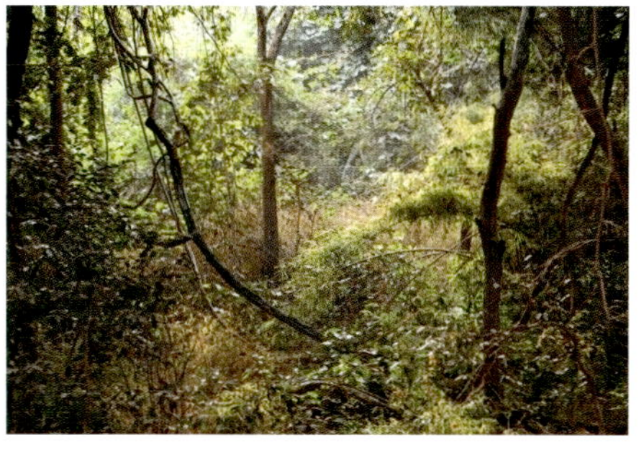

Kanha National Park.

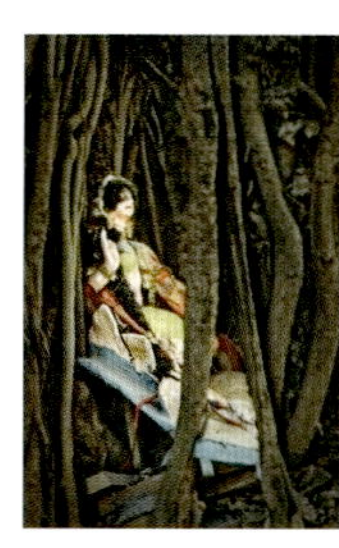

A village shrine along the Brahmaputra River.

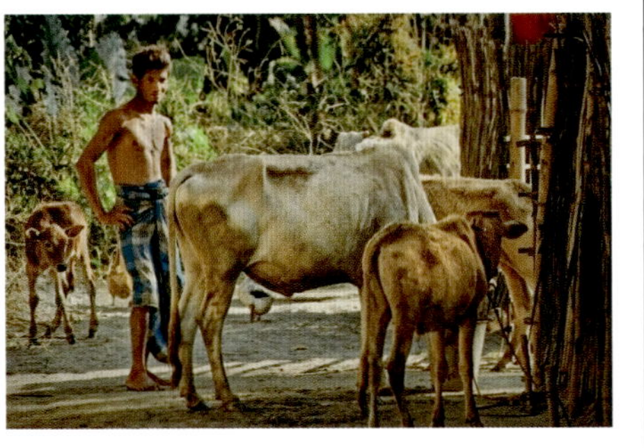

A man and his cows near Guwahati, a major city on the Brahmaputra River.

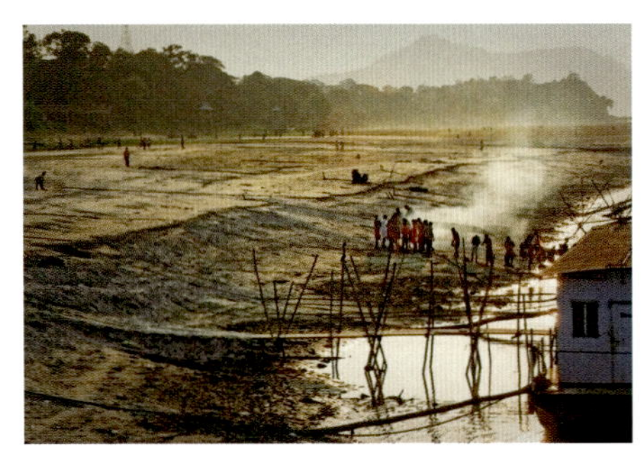

The Brahmaputra River flows some 1,800 miles (2,900 kilometers) from its source high in southwestern Tibet to its confluence with the Ganges River.

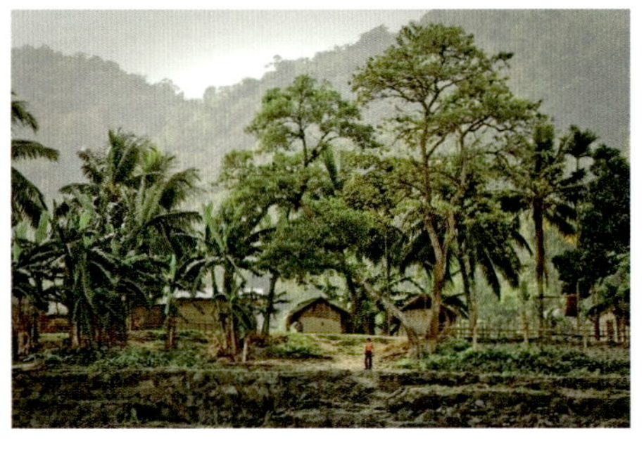

Ganesh Pahar, an idyllically sited village along the Brahmaputra River.

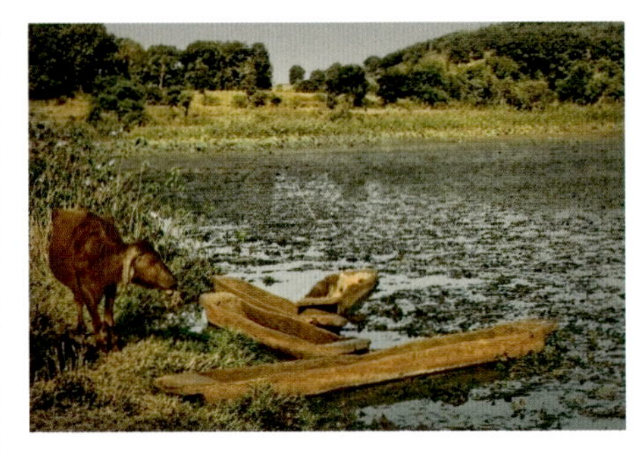

Near Bandhavgarh National Park.

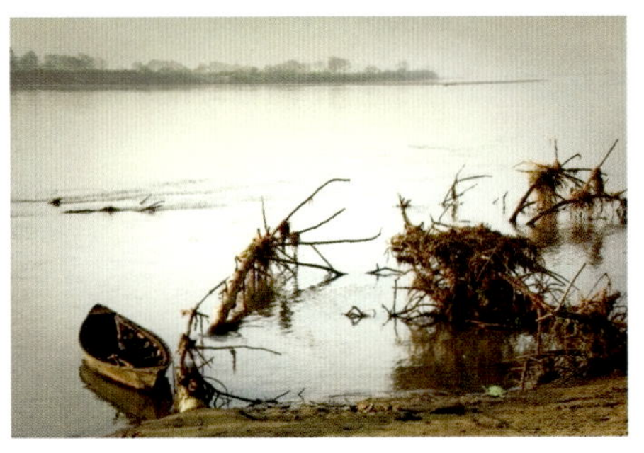

Kaziranga National Park.

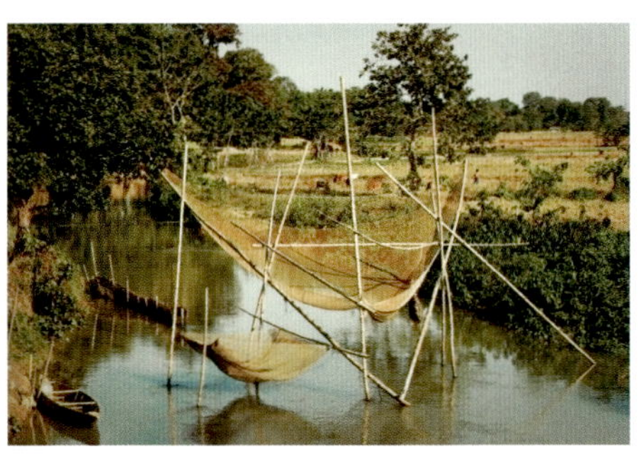

Fish nets near Kaziranga National Park.

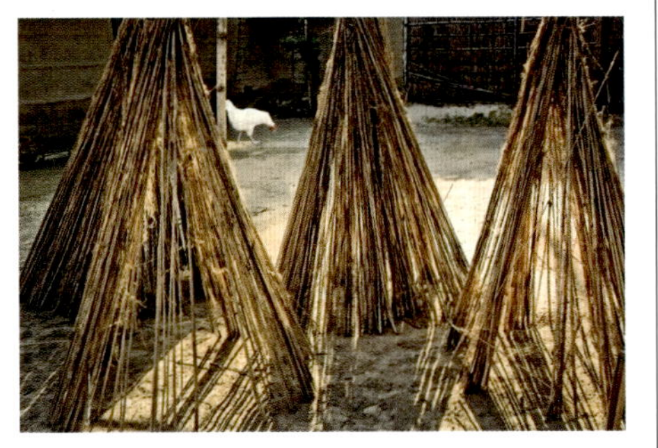

Rural life in Ganesh Pahar.

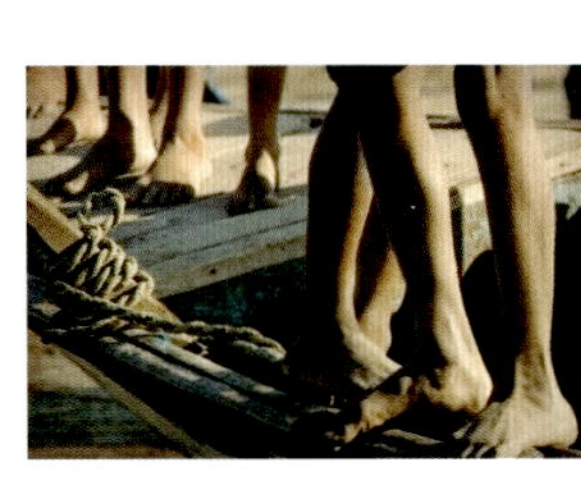

Down by the Brahmaputra River in Ganesh Pahar.

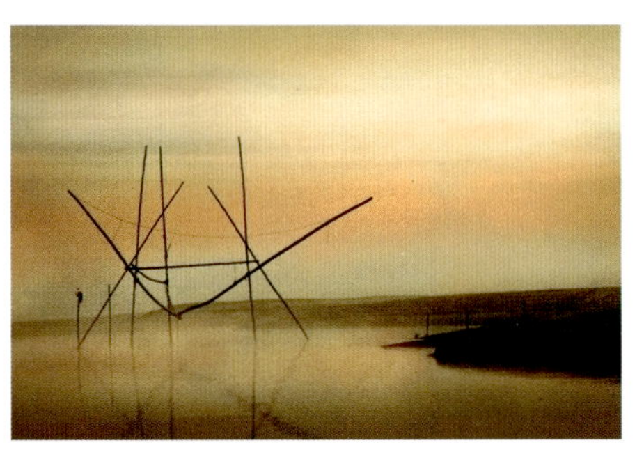

The Brahmaputra River floods during the summer monsoon, creating great, silt-covered floodplains, meadows, and shallow lakes—a variety of luxuriant habitats for birds, reptiles, and mammals.

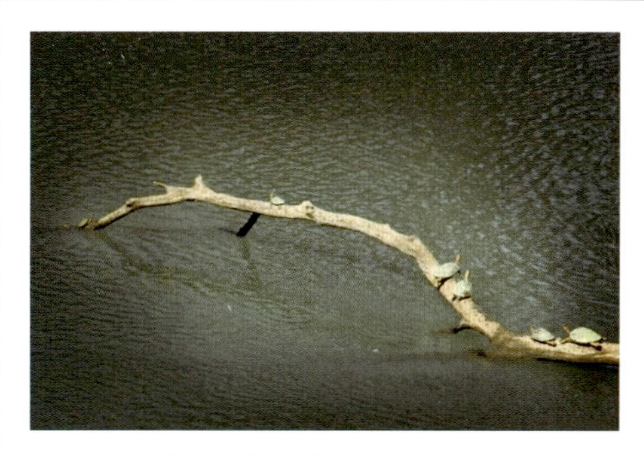

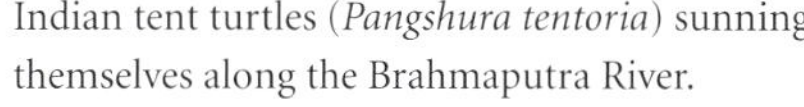

Indian tent turtles (*Pangshura tentoria*) sunning themselves along the Brahmaputra River.

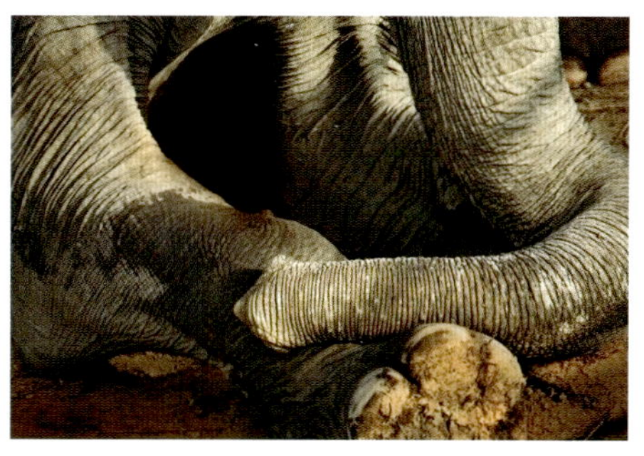

Kaziranga National Park.

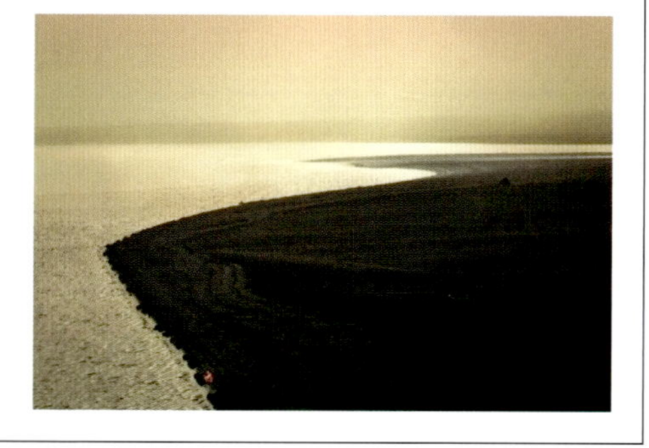

The Brahmaputra River.

About the Author

Joan Myers turned to photography during the early 1970s as her life's work. Her photographs have since appeared in more than fifty solo and eighty group exhibitions throughout the United States, and they are in the permanent collections of the Amon Carter Museum, Bibliothèque Nationale de France, Center for Creative Photography, Denver Art Museum, George Eastman House International Museum of Photography and Film, High Museum of Art, Minneapolis Museum of Art, Museum of Fine Arts, Houston, Museum of Modern Art, Nevada Museum of Art, and San Francisco Museum of Modern Art, among others. Her award-winning books include *Wondrous Cold: An Antarctic Journey* (Smithsonian, 2006), which won an Honorable Mention from the American Association of Museum's 2006 Publications Competition, *Pie Town Woman* (New Mexico, 2001), which was the Best Illustrated Book for 2001 from Publishers Association of the West, *Salt Dreams: Land and Water in Low-Down California,* with William deBuys (New Mexico, 1999), which won both the 1999 Western States Book Award for Nonfiction and the 1999 William P. Clements Prize for the Best Nonfiction Book on Southwestern America, *Whispered Silences: Japanese Americans and World War II* (Washington, 1996), which earned the Rocky Mountain Booksellers Award and an Honorable Mention from Maine Photographic Workshops, *Santiago: Saint of Two Worlds* (New Mexico, 1991), and *Along the Santa Fe Trail* (New Mexico, 1986). Myers lives and works in the village of Tesuque, near Santa Fe, New Mexico.

About the Essayist

Wᴵᴸᴸᴵᴬᴹ ᴅᴇBᵁʸˢ is the author of seven books, including *A Great Aridness: Climate Change and the Future of the American Southwest* (Oxford, 2011), *The Walk* (Trinity, 2007), an excerpt from which won a Pushcart Prize in 2008, *Valles Caldero: A Vision for New Mexico's National Preserve* (Museum of New Mexico, 2006), which won a Southwest Book Award, *Seeing Things Whole: The Essential John Wesley Powell* (Shearwater/Island, 2001), *Salt Dreams: Land Water in Low-Down California,* with Joan Myers (New Mexico, 1999), which won the 1999 Western States Book Award for Nonfiction and the 1999 William P. Clements Prize for the Best Nonfiction Book on Southwestern America, *River of Traps*, with Alex Harris (New Mexico, 1990; Trinity, 2008), which was a *New York Times* Notable Book of the Year and a finalist for the Pulitzer Prize in General Nonfiction in 1991, and *Enchantment and Exploration: The Life and Hard Times of a New Mexico Mountain Range* (New Mexico, 1985), which won a Southwest Book Award and is now in its ninth printing. In 2008, he was awarded a John Simon Guggenheim Memorial Foundation Fellowship in Creative Arts and General Nonfiction. As a long-time conservationist, deBuys has been responcible for protecting more than 150,000 acres of wild Lands in the United States. He also serves on the Board of Directors of the Liz Claiborne Art Ortenberg Foundation, which funds wildlife conservation around the world. DeBuys lives and writes on his farm in El Valle, New Mexico.

Acknowledgements

Many thanks to those who love India's forest and wildlife as I do and who made my little expedition possible, especially Mark Brazil (who also answered many questions) and Zegrahm Expeditions. Thanks to Bill deBuys, for his elegant introduction; he understands that photography is all about seeing, not equipment or f-stops. Without the enthusiasm and considerable efforts of Joanna Hurley, David Skolkin, and George F. Thompson, the book would never have happened nor been so beautiful. And for accompanying me on the journey and putting up with me, *un abrazo* to my husband, Bernie López.

About the Book

The Jungle at the Door: A Glimpse of Wild India was brought to publication in an edition of 2,000 hardcover copies. The text was set in Minion with Mrs Eaves and Avenir display, the paper is Lumi Silk Matte Art, 170 gsm weight, and the book was professionally printed and bound in China but without folios (pagination) for artistic reasons.

The book's title is derived from "Tiger! Tiger!," the third story in *The Jungle Book* (London: Macmillan, 1894), in which Rudyard Kipling wrote: "Most of the tales were about animals, for the jungle was always at their door."; the epigraph (on page 3) is from "Kaa's Hunting," the second story in *The Jungle Book*; and the quotation by William Blake (on page 15) in William deBuys's essay is from "The Tyger," one of twenty-six poems forming the second part of Blake's self-published book, *Songs of Experience* (London: 1794). For bibliographic and other citations, "A Glimpse of the Wild" by William deBuys appears on pages 12–18, "Jungle Journey" by Joan Myers on pages 62–66, and List of Plates on pages 84–91. The book's fifty-six photographic plates begin on page 1 and end on page 83.

Publisher: George F. Thompson
Project Manager: Joanna Hurley, of Hurley Media, L.L.C.
Manuscript Editor: Purna Makaram
Editorial Assistant: Carmen Rose Shenk
Book Design and Production: David Skolkin
Special Acknowledgment: Rudyard Kipling (1865–1936), for his inspirational stories.

George F. Thompson Publishing, L.L.C.
217 Oak Ridge Circle
Staunton, Virginia 24401-3511, U.S.A.
www.gftbooks.com

20 19 18 17 16 15 14 13 12   1 2 3 4 5

The Library of Congress Preassigned Control Number is 2012937160.

ISBN: 978–1–938086–06–9

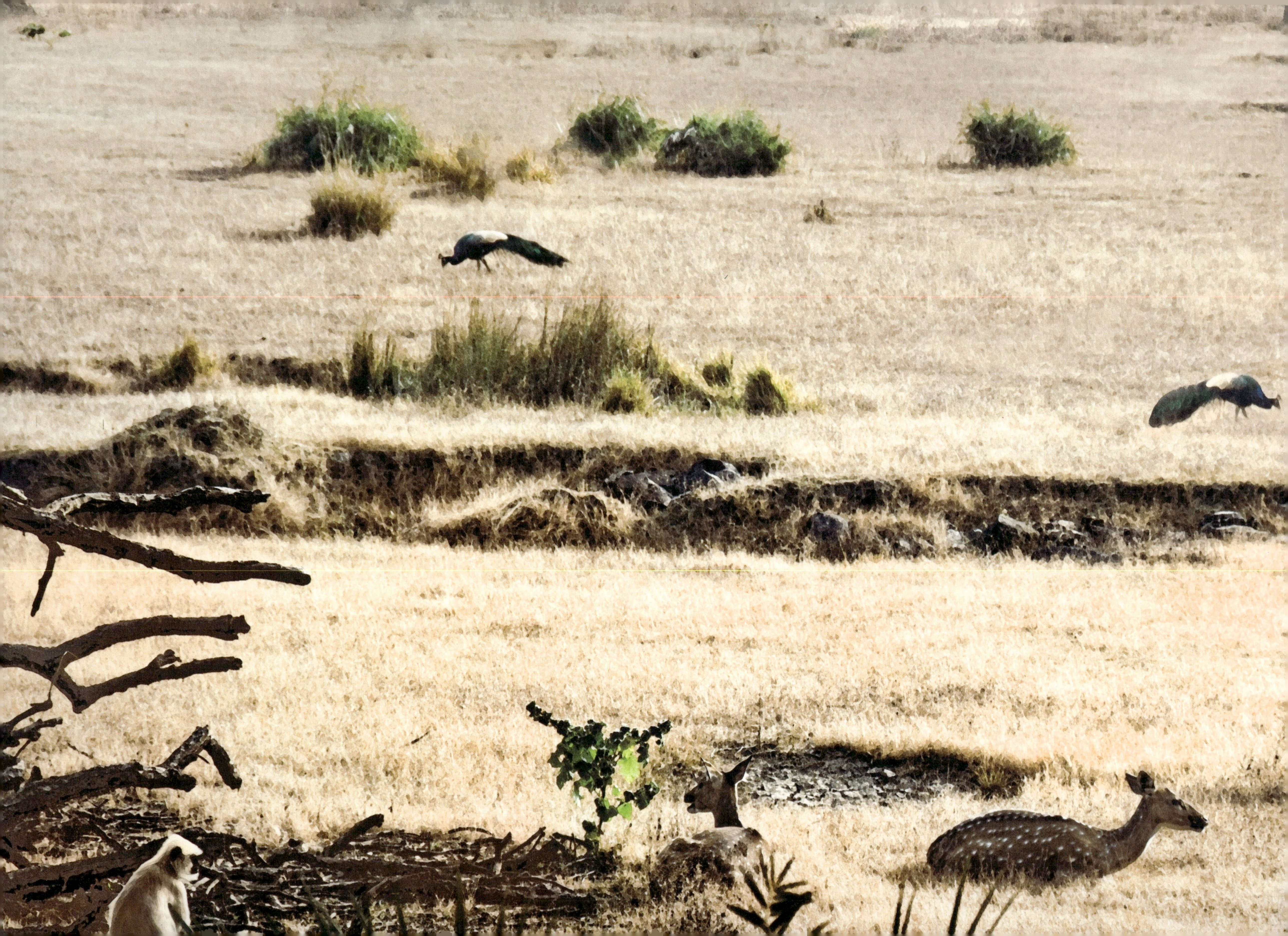